IR Physics 2

The Impact of DLAM on Current State of Scientific Quagmires Holding Us Back

Aaron W. Wemple

CLU & Company

Dear Christian & Corey,

It is with immense joy and gratitude that I dedicate this book to my amazing two sons and team CLU. Your unwavering determination throughout trials in life while writing this book has been invaluable to me. Your passion for the message of Nehemiah, reality-based truth and the connection between the sacred and secular realms has inspired me to share this story with a wider audience.

Your constant faith and trust in God has been an inspiration to us all. Your dedication to using your gifts and talents to bring hope, love and justice to the world is a true testament to the transformative power of God in our lives.

I hope that this book will encourage you, as it has encouraged me, to see the beauty and importance of bridging the sacred-secular divide and to understand that our work, no matter what it may be, is an opportunity to serve God and bring about positive change in the world.

Thank you, Christian and Corey, for being the shining lights in our lives, for inspiring us to be better, and for reminding us that with God, all things are possible.

Blurbs

"Science is not only a disciple of reason but, also, one of romance and passion." - Stephen Hawking

Contents

Preface

Welcome to IR Physics 2 - The Impact of DLAM on Current State of Scientific Quagmires Holding Us Back. This book represents a continuation of our exploration of the fascinating field of IR Physics and the discovery of its innovative concept, DLAM.

As we embark on this journey, we delve into the importance of child safety during times of systemic adult conflict, and the solidification of these principles through the acceptance and validation of others. Our aim is to turn endlessly unrecognizable metrics into recognizable ones by focusing on the "downside-upside" aspects of family, data, and people, and to provide a solid foundation for the longevity of these important rescued issue operations.

This book provides a comprehensive overview of the theories, concepts, and practical applications of IR Physics and DLAM. It aims to defy complacency and push boundaries, leveraging science and technology, creativity, and curiosity to solve the world's problems.

At its core, this book introduces the dynamic leverage of angular momentum (DLAM), which holds the potential to continuously grow in a perpetually positive manner. DLAM can offer solutions to numerous scientific challenges, including perpetual motion, dark energy, the God particle, and free positive power production.

We believe that IR Physics and DLAM are new frontiers in the fields of theoretical physics, life sciences, power production, technology, philosophy, education, and society. We hope that this book serves as a source of inspiration and discovery for readers who seek to understand the potential of this groundbreaking field.

We would like to express our gratitude to all the contributors who have helped to make this book possible, as well as the readers who have taken the time to explore its contents. It is our hope that this book will spark new ideas, generate meaningful discussions, and inspire future research and innovation.

Thank you for joining us on this journey.

Sincerely,

Aaron W. Wemple

Introduction

Introducing the extraordinary world of IR Physics 2! Join us on a journey of discovery as we delve into the fascinating field of IR Physics and its groundbreaking discovery, DLAM. This new edition focuses on the innovations in AI data safety and sustainability, and the potential it holds for the longevity and well-being of individuals and communities.

IR Physics explores the processes of changing form and appearance, and delves into the final state of formation and appearance. It brings forward new concepts, theories, and applications that have the potential to change the world and transform the way we live our lives. IR Physics is a step forward from even classical physics, and has the potential to open your mind to a whole new world of possibilities.

At the core of IR Physics is DLAM, a dynamic leverage of angular momentum that has the potential to continuously grow in a perpetually positive manner. It offers solutions to numerous scientific challenges and has the potential to bring about advancements in clean energy production, science and technology, economy, and society. This field aims to defy complacency and push boundaries, leveraging science and technology, creativity, and curiosity to solve the world's problems.

In this book, we examine the importance of child safety files during times of systemic adult conflict, and the solidification of these principles through the acceptance and validation of others. We also explore the challenges faced during times of systemic conflict and the importance of people's stance on child safety. By focusing on the "downside-upside" aspects of family, data, and people, we aim to turn endlessly unrecognizable metrics into recognizable ones, providing a solid foundation for the longevity of these important rescued issue operations.

IR Physics 2 is a continuous dynamics of perpetually positive forces that explores the inner relational and relevant world, both to our outer methods, motives, and means and to our inner methods, motives, and means of inherent transformation and transfiguration. Finding the correct mutually inclusive transformational degrees of intensity is the key to unlocking this world of limitless potential.

As we embark on this journey of discovery and enlightenment together, let us remember the story of Thomas Edison and the light bulb, which exemplifies the transformative power of IR Physics. Edison's persistence and determination in finding a solution to produce an electric light source revolutionized the world and changed the way people lived their lives. The light bulb was a source of transformative energy that produced positive growth through the complementary laws of attraction, duration, and intensity.

Likewise, IR Physics has the potential to bring about a brighter future for us all, and this book serves as an excellent source of transformative energy that will naturally produce positive growth through the laws of attraction, duration, and intensity. So, join us on this journey of discovery and let's push the boundaries of what's possible.

Chapter One

Overview of IR Physics

In this chapter, we will introduce IR Physics, a new field of study that explores the natural transformation and/or transfiguration of energy. IR Physics aims to understand how energy can be transformed and how this process can lead to advancements in science and technology, as well as benefits to our economy, society, and even our government and charity.

The central concept of IR Physics is the Dynamic Leverage of Angular Momentum (DLAM) Energy Theory, which states that the angular out momentum in rotating networks can be different from the angular momentum in systems. This difference can be adaptively harnessed and transformed into positive energy growth for all.

The implications of DLAM Energy Theory are far-reaching and can impact not just the field of physics but also power production, life sciences, technology, philosophy, economy, education, and society as a whole.

In this chapter, we will provide a brief overview of the DLAM Energy Theory and its applications, including:

- Power innovation implications: How DLAM Energy Theory can be applied to produce clean and renewable energy, solve problems such as perpetual motion, dark energy, and the God particle.

- Social innovation implications: How DLAM Energy Theory can be used to benefit society and improve our economy, government, and charity.

- Academic innovation implications: How DLAM Energy Theory can be used to advance our understanding of the world and push the boundaries of science and technology.

We hope that this overview will give you a better understanding of IR Physics and the potential it holds for changing our world. In the coming chapters, we will delve deeper into the theory and its applications, providing a comprehensive and in-depth look at IR Physics.

"Angular momentum is like a spinning top, twirling and dancing to its own rhythm. It's a force to be reckoned with, never losing its energy unless something comes along to give it a nudge. But with DLAM, we're talking about a spinning top on steroids! It's a powerhouse, driving itself in every direction imaginable. Think of it like the ultimate spinning top that never stops. The secret to unlocking the full potential of both angular momentum and DLAM is in mastering the art of directing its changes. It's like being a spinning top conductor, leading the dance with a flick of the wrist."

Power innovation implications

The study of Dynamic Leverage of Angular Momentum (DLAM) Energy Theory has the potential to revolutionize the production of clean and renewable energy sources. By exploring the changes in angular momentum, this theory aims to address long-standing scientific quandaries such as perpetual motion, dark energy, and the elusive God particle. The key to unlocking the power of DLAM lies in understanding the ways in which these changes can be harnessed and applied in practical, energy-producing systems.

Similar to perpetual motion experiments, the exploration of DLAM has involved experimentation and observation to determine the ways in which its principles can be harnessed and applied to real-world problems. Scientists and researchers have conducted various experiments to understand how the dynamic leverage of angular momentum can be leveraged to produce clean and renewable energy, solve challenges in physics such as perpetual motion, dark energy, and the elusive "God particle," and more. The key to unlocking the potential of DLAM lies in understanding the nature of its changes and harnessing its energy through careful experimentation and observation. By studying the dynamics of angular momentum and its changes, we can gain a deeper understanding of the potential for DLAM to provide new solutions to old problems and drive innovation in a variety of fields.

When a ball is kicked, it absorbs the impact from the foot that kicked it and begins to curve in its trajectory. This curving motion is due to the combined effect of multiple forces, including the force of the kick, the force of gravity, and the force of air resistance. These forces all act upon the ball, changing its angular momentum in various directions and leading to the ball's curved path.

Similarly, when one side of a planet is facing the hot sun, that side may begin to spin differently than the cool side. This difference in rotation is due to the compounding of multiple forces, including the force of the sun's heat, the force of the planet's gravity, and the force of the planet's magnetic field. These forces all act upon the planet, changing its angular momentum in various directions and leading to differences in the speed and direction of rotation between the hot and cool sides of the planet.

By understanding the compounding of angular momentum and how it affects objects in motion, we can gain a deeper understanding of the dynamic changes that occur in the natural world. This knowledge can then be applied to fields such as energy production and space exploration, where the principles of angular momentum and DLAM play a critical role.

DLAM theory posits that the expansion and compression sides of a two-dimensional rotating sphere are affected differently by the "going out" and "going in" angular momentum. According to DLAM, the "going out" angular momentum on the expansion side results in a different outcome compared to the "going in" angular momentum on the compression side. This difference in behavior challenges the classical theory of energy conservation, which states that energy cannot be created or destroyed but can only change form. The DLAM theory suggests that the conservation of energy may not always hold true in certain rotating systems. By taking into account the differences in angular momentum between the expansion and compression sides, DLAM provides a new perspective on the behavior of rotating systems and could have far-reaching implications for our understanding of energy and its conservation.

Imagine you're at a theme park, riding a giant spinning teacup! The simple angular momentum of this ride is like a spinning top, it keeps going round and round in one direction, unless acted upon by an outside force, such as a bump from another teacup or the rider's hand trying to stop the spinning.

Now, imagine you're on a different ride, a DLAM-powered teacup! This teacup is not just spinning in one direction, but it's constantly adjusting and changing its spin direction, almost like it has a mind of its own. This is because the DLAM energy on the teacup can drive itself in all directions, like a wild, untamed beast. The ride is much more unpredictable, with twists, turns, and unexpected changes in direction.

So, while the simple angular momentum teacup is like a predictable spinning top, the DLAM teacup is like a chaotic, wild ride. The key difference is that the DLAM teacup is able to harness the changes in its angular momentum to create an exciting and unpredictable experience.

Social innovation implications

The application of DLAM Energy Theory has the potential to significantly impact various aspects of society and bring about a positive change. This can include economic growth, more efficient government systems, and improvements in charity and philanthropic efforts. The utilization of DLAM energy has the power to drive social innovation and bring about a brighter future for all.

An example of the social innovation implications of DLAM Energy Theory can be seen in the application of this theory to improve the energy efficiency of our homes and businesses. By harnessing the power of DLAM to capture and use energy from natural sources such as wind and sun, we can reduce our dependence on fossil fuels and improve our environment. Furthermore, DLAM Energy Theory can be used to improve the way our economy operates by reducing energy costs and increasing the production of clean, renewable energy. Additionally, DLAM Energy Theory can be used to help improve government by providing a new source of revenue, as well as by reducing the need for government subsidies and support of the fossil fuel industry. To test these implications of DLAM Energy Theory, we could conduct case studies to see how DLAM Energy Technology has improved the energy efficiency of homes and businesses, analyze the economic impact of DLAM Energy Technology on the energy market, and examine the political and social impact of DLAM Energy Technology on government policies and public opinion.

"DLAM Energy Theory is not just about harnessing power, it's about harnessing the power of change for the betterment of society. Think of it as a turbo boost for the heart and soul of our communities!"

Academic innovation implications

DLAM Energy Theory has the potential to be a catalyst for revolutionary change in the academic world. Challenging long-held beliefs and expanding the boundaries of our understanding, can open up new avenues of exploration and discovery, leading to major advancements in science and technology.

"With DLAM Energy Theory, the sky's the limit for academic innovation! Imagine unlocking secrets of the universe, from the tiniest particles to the vast expanse of space. With DLAM, the impossible becomes possible and the unknown becomes known!"

Once upon a time, there was a brilliant physicist named Dr. Elena who had a burning passion for discovering new and innovative theories that could change the world. She spent countless days and nights researching and experimenting until one day, she stumbled upon something that would revolutionize the field of physics forever: IR Physics.

Dr. Elena's new field of study explored the natural transformation and transfiguration of energy, which she believed could lead to advancements in science, technology, and benefit society as a whole. Her research led her to discover the central concept of IR Physics: the Dynamic Leverage of Angular Momentum (DLAM) Energy Theory.

Through her research, Dr. Elena discovered that the difference in angular momentum between rotating networks and systems could be harnessed and transformed into positive energy growth. The implications of DLAM Energy Theory were vast, and it could impact not just the field of physics but also power production, life sciences, technology, philosophy, economy, education, and society as a whole.

Dr. Elena was ecstatic about the potential of her discovery and wrote a book about it, providing a brief overview of the DLAM Energy Theory and its applications. In her book, she discussed the power innovation implications of her theory, such as producing clean and renewable energy, solving problems like perpetual motion, dark energy, and the God particle.

Furthermore, Dr. Elena also discussed the social innovation implications of DLAM Energy Theory, how it could benefit society and improve our economy, government, and charity. She was also excited to explore the academic innovation implications of her theory, how it could advance our understanding of the world and push the boundaries of science and technology.

Dr. Elena hoped that her book would give people a better understanding of IR Physics and the potential it holds for changing our world. She continued to delve deeper into the theory and its applications, providing a comprehensive and in-depth look at IR Physics. Her work inspired a new generation of physicists to continue exploring the natural transformation and transfiguration of energy for the betterment of society.

In chapter one, we discussed the Power, Social, and Academic innovation implications of DLAM Energy Theory. We explored how DLAM can be used to produce clean and renewable energy, benefit society and improve our economy, government, and charity, as well as advance our understanding of the world and push the boundaries of science and technology.

Now, in chapter two, we will delve deeper into the theory behind DLAM and examine the concept of transformation and transfiguration. We will explore how DLAM's unique ability to harness the changes in angular momentum can lead to truly game-changing innovations and advancements. So buckle up, it's time to go on a journey of discovery and unlock the secrets of DLAM!

Chapter Two

The Concept of Transformation and Transfiguration

"Why did the physicist transfigure into a mathematician? Because she wanted to make transformation even more complex!"

In this chapter, we will delve deeper into the concept of transformation and transfiguration as they relate to DLAM Energy Theory. The term "transformation" refers to the process of changing from one form or state to another, while "transfiguration" refers to a complete change of form or appearance, often accompanied by a spiritual transformation.

Do you know how a caterpillar changes into a butterfly? That is an example of transformation. Transformation is when something changes from one form or state to another.

Transformation is a term used to describe the process of changing from one form or state to another. It is a concept that is widely used in various fields, including physics, chemistry, and biology. In physics, for example, transformation is often used to describe the change in energy or matter from one form to another, such as the conversion of potential energy to kinetic energy. In chemistry, transformation is often used to describe chemical reactions, such as the conversion of water to hydrogen and oxygen. Overall, transformation is a fundamental concept that helps us understand how various systems in the world around us change and evolve over time.

Once upon a time, there was a caterpillar named Charlie who lived a simple life crawling on leaves and munching on them all day. One day, Charlie decided he wanted something more and began to feel a strange feeling inside him. Suddenly, he spun a cocoon around himself and transformed into a butterfly.

At first, Charlie was amazed at his new form and the beauty of his wings. He flapped his wings and took off into the sky, feeling the wind beneath him. But as he flew, he noticed that something was different. He felt like he had transformed not just his body, but also his mind, his soul, and his essence. He was no longer just a butterfly, but a completely different creature. He had undergone transformation.

From that day on, Charlie saw the world in a completely new light. He could soar higher, fly faster, and see things from a completely different perspective. He had transcended his old form and become something greater.

Just like Charlie the caterpillar, DLAM theory can also bring about a transformation in the way we think about energy, our world, and the universe. By exploring and harnessing the power of dynamic leverage of angular momentum, we can achieve great breakthroughs in science and technology and bring about positive change in society.

Transfiguration means when something changes its form or looks completely different, sometimes because of a spiritual change. For example, let's say you have a plain rock named Rocky. If Rocky suddenly turned into a beautiful flower named Daisy, that would be transfiguration.

Transfiguration is the process of a complete change of form or appearance, often accompanied by a spiritual transformation. This term is often used in religious contexts to describe the transformation of Jesus Christ before his disciples. In scientific terms, transfiguration can be observed in the transformation of matter, such as the conversion of ice to water or water to steam. It is also used in literature and art to depict profound personal transformation, as in the character Ebenezer Scrooge in Charles Dickens's "A Christmas Carol," who experiences a transfiguration of his personality after being visited by three ghosts.

We will explore how DLAM Energy Theory can be applied to bring about transformation and transfiguration in various areas of society. For example, DLAM Energy Theory has the potential to transform our energy systems and reduce our reliance on non-renewable sources of energy. This, in turn, can bring about a transfiguration in our approach to sustainability and the environment.

Imagine a small town struggling with poverty and income inequality. The local government has tried various solutions, but nothing seems to work. That's when a group of DLAM applied scientists come to town, armed with the knowledge of DLAM Energy Theory. They work with the government and local businesses to harness the power of DLAM and create new and innovative solutions to the problems facing the town. Before long, the town's economy is thriving, poverty is reduced, and income inequality is greatly reduced. The transformation and transfiguration brought about by DLAM Energy Theory have created a better future for everyone in the town.

Similarly, DLAM Energy Theory can bring about transformation and transfiguration in the fields of economics, government, and charity. By leveraging the power of DLAM, we can create new and innovative solutions to the problems that face society, such as income inequality and poverty.

Imagine a small organization where poverty and income inequality are rampant. The traditional methods of solving these problems, such as distributing wealth through government programs or charitable donations, have not been effective in bringing about lasting change.

Enter DLAM Energy Theory. By harnessing the power of DLAM, local entrepreneurs are able to create new and innovative solutions to poverty and income inequality. They develop small-scale, sustainable energy sources that provide affordable electricity to the community. This not only reduces the cost of living but also creates new job opportunities, driving economic growth.

Example: A local entrepreneur in a rural area creates a new type of micro-grid system that utilizes DLAM energy. This system provides low-cost electricity to families in the community, empowering them to start small businesses and lift themselves out of poverty. The entrepreneur partners with local townships to provide the technology at a reduced cost to the most vulnerable families in the community, and works with local schools to educate the next generation on how to maintain the system.

In addition, the entrepreneurs work with local government officials to implement new economic policies that encourage growth and distribute wealth more equally. They also partner with local charities to provide education and job training programs to those in need, empowering them to take control of their own lives and break the cycle of poverty.

Through the transformative power of DLAM Energy Theory, the small town is able to make a significant impact on the lives of its residents and provide a model for other communities to follow.

Example 2: Another local entrepreneur creates a new type of agricultural system that utilizes DLAM energy to reduce the cost of irrigation. This system increases crop yields and reduces the cost of food for families in the community. The entrepreneur partners with local government officials to implement new economic policies that encourage the growth of small-scale agriculture, and works with local banks to provide loans to farmers who want to adopt the new technology. The result is a more sustainable and resilient local economy, where entrepreneurs, government, and banks work together to create a brighter future for the community.

Moreover, DLAM Energy Theory has the potential to push the boundaries of science and technology, leading to a transformation in our understanding of the world and our place within it. This, in turn, has the potential to drive technological innovations and advancements in a range of fields, from medicine to transportation.

Example 3 of how DLAM Energy Theory can push the boundaries of science and technology and lead to a transfiguration in our understanding of the world and our place within it, could be seen in the field of data analysis. By using DLAM to analyze data metrics, we can not only create more accurate and precise data models, but also design data metrics that are geared specifically for each next generation. This would be just as important as policy metrics that are created and modified for each passing generation. With DLAM, data analysis becomes not just a way to track trends and patterns, but a tool for shaping the future and promoting innovation in a range of fields, from medicine to transportation.

In conclusion, the concept of transformation and transfiguration is a central theme in DLAM Energy Theory, and in this chapter, we will explore its implications for society, the environment, and the advancement of science and technology.

Why did the old, mean boundary cross the road?

To get to the civilized side, where it could be a little more fair and friendly!

Chapter Three

The Significance of the Dynamic Leverage of Angular Momentum (DLAM)

In recent years, the field of physics has made significant strides in understanding the behavior and properties of matter and energy. One of the most exciting new developments in this area is the concept of Dynamic Leverage of Angular Momentum, or DLAM.

The Dynamic Leverage of Angular Momentum (DLAM) is different from angular momentum in that DLAM takes into account the changing distribution of mass in a rotating object, whereas angular momentum only considers the object's mass and velocity.

DLAM is a fancy way to talk about how things spin around. Imagine you and your friends are playing with spinning tops. When you spin them, they go around and around, right? But sometimes they might start wobbling, spinning slower, or bouncing off each other. That's because the top gets heavier on one side and lighter on the other side, and that makes it harder to keep spinning. DLAM helps us understand why that happens!

DLAM is a concept in physics that describes the relationship between the distribution of mass and the rotational motion of interacting objects in motion. When objects rotate, the masses are not distributed evenly, which may or may not affect every other rotational behavior. DLAM takes into account the changing internal and external network of mass and the resulting changes in the object's rotational motion.

For example, imagine a figure skater spinning on the ice. As the skater pulls their arms in closer to their body, their mass becomes more concentrated towards the center of their body. This changes the distribution of mass in the skater-object system, which affects their rotational behavior. DLAM helps us understand how these changes in mass distribution affect the skater's spinning motion.

The Dynamic Leverage of Angular Momentum (DLAM) is different from the traditional concept of conservation of energy in a number of ways. While the conservation of energy theory states that energy cannot be created or destroyed, only transferred from one form to another, DLAM focuses on the behavior of angular momentum in rotating systems, specifically how it changes and how it can be harnessed. Additionally, the conservation of energy theory applies to isolated systems, whereas DLAM applies to open systems that are subject to duo outside and inside forces. This distinction allows for the study of the effects of duo forces on angular momentum and how those forces can be leveraged to create new forms of energy and drive innovation in various fields.

DLAM (Dynamic Leverage of Angular Momentum) and dark energy are two distinct concepts in physics and cosmology. Dark energy is a hypothetical form of energy that is thought to exist in the universe and to be responsible for the observed accelerated expansion. It is often described as a fluid or field that fills the universe and has negative pressure, causing the expansion to accelerate.

In contrast, DLAM is a theory that explains the dynamics of rotating spheres and the potential sharing of energy in such networks. It proposes that the flexing of angular momentum can lead to the creation of new energy, which in turn can be harnessed for practical applications, such as generating clean and renewable energy.

Therefore, DLAM and dark energy differ in their scope, origins, and implications. DLAM focuses on the energy created through the transfer of angular momentum in rotating spheres, while dark energy pertains to the overall energy of the universe and its influence on the rate of expansion.

Max and Dr. Liana led a team of figure skaters who were preparing for an important competition. The skaters had been practicing for weeks, but Max noticed that their performances lacked that special "wow" factor.

One day, during practice, Max decided to explain to the team the difference between regular angular momentum and DLAM.

"Okay, guys, so here's the deal," Max said. "Regular angular momentum is like when you spin on one leg. You use your mass and velocity to keep you spinning. But with DLAM, you're not just spinning on your own. You're working as a team, and the changing distribution of mass can help you do amazing things!"

The skaters looked at Max with blank stares. Dr. Liana stepped in. "Think of it like this," she said. "When you spin on your own, you're like a soloist. You can do some pretty cool tricks, but there's only so much you can do. But when you work as a team, you're like an orchestra. You can create something truly magnificent."

Suddenly, the skaters understood. They began to work together in new ways, using the principles of DLAM to create breathtaking performances. Lifts and spins that had seemed impossible before now seemed effortless.

At the competition, the skaters wowed the judges with their dynamic performances. They knew that they had learned something important about teamwork and the power of DLAM.

Max and Dr. Liana couldn't have been prouder of their team. They knew that they had helped them unlock their full potential. And who knows? Maybe they had even inspired the judges to think about the physics of figure skating in a whole new way!

Mark loved motocross racing and was also an engineering student. One day, he was on the track and wondered why conventional physics couldn't explain a lot of things that he and other riders saw while flying wildly through the air and curves. He realized that regular angular momentum did not apply to their fast bikes and that something was missing from conventional physics.

After school, Mark conducted experiments and discovered that the changing distribution of mass in the bike and the rider's body position contributed to the bike's movement. He called this phenomenon the Dynamic Leverage of Angular Momentum or DLAM. With this new understanding, Mark and his motocross team began to implement DLAM in their racing strategy.

During their races, Mark and his teammate would use their internal bodies to navigate the bike through the curves and jumps. They would shift their weight to maintain balance, and their movements would affect the bike's rotation and momentum. Their improved duo navigations with the external circumstances of the track resulted in faster lap times and more successful races.

Mark's discovery of DLAM revolutionized the way motocross racers approached their sport. It allowed them to better understand the physics behind their movements, resulting in more efficient and effective racing strategies. Thanks to Mark's curiosity and innovation, DLAM became an essential tool for motocross racers worldwide.

It may seem unrelated, but Max and Mark's experiments on DLAM could potentially shed light on the mysteries of dark energy and dark matter. By studying the behavior of rotating objects with changing mass distributions, they could uncover new principles of physics that could eventually lead to a unifying theory that explains both dark energy and dark matter. For example, they may discover new fundamental forces or particles that interact with these mysterious phenomena in ways that we haven't yet observed. While it may be a long shot, any new insights into the nature of the universe could potentially have profound implications for our understanding of dark energy and dark matter.

Dark energy and dark matter are mysterious phenomena that constitute a large portion(95%) of the observable universe, but their physical nature is still not well understood. The lack of a unifying theory that combines both of these phenomena has been a challenge for physicists.

DLAM, on the other hand, is a theory that focuses on the dynamic leverage of angular momentum, which is distinct from dark energy and dark matter. DLAM suggests that angular momentum can be used to explain the behavior of objects in motion and that the interaction between the "going in" and "going out" angular momenta can impact the system differently than classical conservation of energy.

DLAM is a theory that focuses on the dynamics of angular momentum, while dark energy and dark matter focus on the mysterious phenomena that make up a large portion of the observable universe. The two concepts are distinct and address different areas of study.

A relatively new concept of explaining the unexplained is parallel universe theory. It is based on the idea that there are multiple universes, existing alongside our own, with different physical constants and properties. The multiverse theory is still purely speculative and has not been proven through empirical evidence.

In contrast, DLAM Energy Theory is based on complimenting the established laws of physics and has been demonstrated through experiments and simulations. The theory posits that the Dynamic Leverage of Angular Momentum (DLAM) is a new form of energy that can be harnessed and applied in practical ways to benefit society. Unlike the idea of parallel universes, DLAM Energy Theory has a solid foundation in science and has the potential to be applied in a range of fields, from energy production to transportation.

In summary, while the concept of parallel universes is an interesting idea, it remains purely speculative and has limited practical applications. DLAM Energy Theory, on the other hand, is grounded in science and has the potential to bring about real-world benefits and innovations.

DLAM represents a fundamental shift in our understanding of energy and its role in shaping the world around us. This innovative theory provides a new framework for understanding how matter and energy interact, and how we can harness the power of these interactions to drive technological advancements and address pressing societal challenges.

The significance of DLAM lies in its ability to leverage the energy of rotating objects in a dynamic manner, creating new and innovative solutions to problems in fields ranging from energy production and transportation to economics and charity.

Imagine you're sitting inside a shopping cart at the grocery store, gently rocking back and forth. Now imagine that same experience, but instead of being in a store, you're floating in outer space. The cart is now spinning and you're feeling the centrifugal force pulling you against the sides. Suddenly, a meteor hits your cart and sends it careening off in a new direction. This is where DLAM comes in.

DLAM, or the Dynamic Leverage of Angular Momentum, is about much more than just rocking back and forth in a shopping cart. It's about leveraging the energy of rotating objects in a dynamic and controllable manner. Instead of being at the mercy of every little bump or meteor that comes your way, with DLAM you could use the energy of your spinning cart to steer yourself to safety. The significance of DLAM lies in its ability to harness and harness the energy of rotating objects dynamically, not just to sit in a shopping cart and go places inside a store. So while parallel universes might be a fun idea, DLAM has practical applications that can be leveraged in a variety of fields.

This is where DLAM Energy Theory comes into play. By leveraging the energy of rotating objects, such as the rocket, DLAM has the potential to increase the efficiency and effectiveness of the deflection mission. Instead of relying solely on brute force, DLAM allows for a more dynamic and nuanced approach, harnessing the energy of rotation to achieve the desired outcome. This could be game-changing for the field of space exploration and meteor deflection, potentially providing a more efficient and effective way to protect our planet from dangerous incoming objects.

Imagine you're surfing on a light wave traveling through outer space. You're riding the wave, feeling the energy and momentum pushing you forward. Suddenly, you encounter a massive black hole, and the gravitational pull of the black hole disrupts your surfing experience. You feel yourself getting pulled in and you need to adjust your balance and positioning to compensate for the change in gravitational force. This is where DLAM comes in. By taking into account the changing distribution of mass in the black hole and the surrounding space, DLAM can help you better understand and navigate the gravitational forces impacting your surfing experience on the light wave.

As we continue to explore and understand the implications of DLAM, it is clear that this theory has the potential to transform the way we think about energy and its role in shaping our world. Whether we are looking to create new sources of clean, renewable energy, or to address pressing issues such as poverty and income inequality, DLAM represents a powerful tool for advancing our knowledge and creating a better future for all.

As we delve further into the realm of DLAM, we come across a new frontier in its application and potential. In Chapter Four, we explore the cutting-edge breakthroughs in the field of Quasi-Kinetically Charged Dynamic Leveraged Angular Momentum. This exciting new development promises to take DLAM to even greater heights, unlocking its full potential to drive innovation and bring about transformation in a range of fields. So buckle up and get ready to embark on a journey of discovery as we delve into the world of supercharged DLAM.

Chapter Four

Advances in Quasi-Kinetically Charged Dynamic Leveraged Angular Momentum (DLAM)

In chapter three, we delved into the significance of Dynamic Leveraged Angular Momentum (DLAM) and its potential to revolutionize a range of fields, from power production to sports. Now, in chapter four, we delve into the groundbreaking advances in quasi-kinetically charged DLAM.

Building upon the principles of conservation of energy, DLAM takes it a step further by incorporating the transfer of duo angular momentums and their effect on the energy of a system. This results in a more comprehensive understanding of the relationship between energy and its transfer.

The conservation of energy principle states that the total energy within a system is constant, and can only be transferred, not created or destroyed. This can be mathematically expressed as the equation $\Delta E = Q + W$, where ΔE represents the change in total energy, Q represents heat energy transfer, and W represents work done on or by the system. The net energy gain or loss within the system can be determined by examining the value of ΔE; if it is positive, there is a net energy gain, and if it is negative, there is a net energy loss. A net energy gain example could be a solar panel converting sunlight into electrical energy, as the total energy within the system increases. A net energy loss example could be a light bulb producing heat energy, as some of the total energy within the system is lost to the surroundings.

It is important to note that this equation only pertains to internal energy changes and transfers, and does not account for external factors that may affect the energy balance of a system. For example, if you're trying to determine the energy balance of a pot of water on a stove, the equation would only consider the heat energy transfer and work done within the pot itself, and would not account for any heat lost to the surrounding environment due to convection or radiation.

There are certain types of physical and chemical systems known to violate the principle of conservation of energy. These systems are known as open systems, which are characterized by the exchange of energy and matter with their surroundings. Examples of open systems include living organisms, thermodynamic systems, and systems involving friction or non-conservative forces. In these systems, the internal energy can change due to heat transfer, work done on the system, and energy exchange with the environment. Hence, the conservation of energy principle is not always applicable in open systems, as energy can be gained or lost through the exchange with the surroundings.

Imagine playing with a balloon. When you blow air into it, it gets bigger and has more energy. But when you let the air out, the balloon gets smaller and loses energy. This is like open systems where energy can be gained or lost through exchange with the surroundings.

Open systems are physical or chemical systems that can exchange matter or energy with their surroundings. Examples of open systems include biological organisms and thermodynamic systems. These systems can violate the conservation of energy principle as their internal energy can change due to energy exchange with the environment. The exchange can occur through heat transfer, work done on the system, or non-conservative forces such as friction. As a result, the conservation of energy principle is not universally applicable to open systems.

Kinetic energy is the energy an object possesses due to its motion.

Imagine throwing a ball. The harder you throw it, the more energy it has, and the farther it will go. That's kinetic energy.

Kinetic energy is a scalar quantity that depends on both the mass and velocity of an object. The equation for kinetic energy is $KE = 1/2mv^2$, where m is the mass of the object and v is its velocity.

In some cases, the boundary between two open systems can create what's called quasi-kinetic energies. These energies arise due to the transfer of mass, heat, or momentum between the two systems and can be difficult to quantify because they don't fit neatly into the traditional definitions of kinetic or potential energy. An example of this might be the exchange of energy between the Earth's atmosphere and the ocean, or between the human body and the surrounding environment during physical exercise.

Quasi-kinetic energies occur at the boundaries where open systems interact with each other. For example, consider two open systems: a hot cup of coffee and a room-temperature mug. When the hot coffee is poured into the room-temperature mug, energy is transferred from the hot coffee to the mug, which results in a decrease in the temperature of the coffee and an increase in the temperature of the mug. This transfer of energy is known as quasi-kinetic energy because it occurs at the quasi-boundary between the two open systems. In this example, the heat energy of the hot coffee acts as the quasi-kinetic energy, and the transfer of energy from the coffee to the mug represents a violation of the conservation of energy principle, as the total internal energy of the two systems has changed.

An example of where quasi-kinetic energy borders in rotation meet each other could be in a planetary system, where two celestial bodies such as two planets are rotating around a common center of mass, and their orbits intersect. The energy borders at the intersection points create an exchange of energy between the two bodies, resulting in the transfer of kinetic energy. This transfer of energy results in a change in the rotational velocity and momentum of the bodies, leading to a rearrangement of their orbits. The concept of quasi-kinetic energy transfer at the borders of interacting rotating systems can be described by the DLAM halo effects.

DLAM halo effects occur when there is a transfer of quasi-kinetic energy at the intersection points of rotating systems, such as in a planetary system where two planets are rotating around a common center of mass, and their orbits intersect. As a result of the exchange of energy, the rotational velocity and momentum of the planets can be altered, leading to a rearrangement of their orbits. For example, the transfer of energy between two planets can cause one planet to be pulled closer to the sun, while the other is pushed farther away, resulting in changes to their orbits and the overall dynamics of the system.

As of the current knowledge cut-off, there is not yet an equation for DLAM because the theory is still in its developmental stage. The concept of DLAM is still being explored and studied, and more research is needed in order to fully understand and formulate an equation that accurately describes the phenomenon. Currently, DLAM is mainly based on empirical observations and qualitative descriptions, and a formal mathematical description is yet to be developed. This is a common occurrence in the early stages of scientific theories, as further investigation and experimentation are required in order to develop a more precise and comprehensive understanding of the underlying phenomena.

If one had to assume an equation for DLAM, each 'pocket' of its existence is custom for that situation. Thus, it would be difficult to develop a general equation for DLAM that could encompass all of its possible expressions. Additionally, as the concept of DLAM is still a relatively new idea in the field of physics, much more research and experimentation are needed to fully understand its behavior and properties before a comprehensive equation can be derived.

After thorough applications analysis, it is clear that IR Physics theories have provided significant contributions to the advancements in social, economic, and technological fields. One such application is the use of DLAM in constructive child safety file metrics during high conflict divorce cases, adult fighting file trial metrics, and similar contract disputes. DLAM has proven to reveal real-time times in the midst of exaggerated overlapping hostilities, leading to new legislative initiatives such as Christian's Law, Corey's Act, Madison's Initiative, Evan's Initiative, and Alayna's Law. These initiatives aim to improve high conflict divorce pipeline circumstances, create a national reuniting party, limit the separation time between children and their parents due to litigation continuances, provide bond allowances for victims of false accusations in divorce court, and deter children from being denied phone contact with both parents following divorce deals.

It is worth noting, however, that not everyone can have a super, quasi-angelic life like all the benefactors in this book, and the implications of DLAM on power innovation are yet to be fully explored.

It is crucial that we urgently recognize the need for childproofing often psychologically damaging deals that American children are forced to endure in conflict situations. The application of DLAM in the realm of divine and SocioEcTech has proven to have a significant impact in areas such as child custody and adult conflict resolution. DLAM's unique ability to provide real-time metrics in the midst of overlapping lag-time hostilities is crucial in resolving contract disputes and high-conflict divorce cases. We must empathize with the suffering of these children and take action to support legislation such as Christian's Law, Corey's Act, Madison's Initiative, Evan's Initiative, and Alayna's Law, as well as the emergence of new startups in this field.

God's divine plan may not always be apparent to us, but we must remember that He is the mastermind behind everything in existence. It's crucial to acknowledge the blessings that come our way, such as new knowledge and understanding, as they are a testament to His goodness and grace. We should be grateful for the opportunities He provides us to learn and grow, even when we face challenges that may obscure our view of His plan.

As Jesus teaches in Matthew 5:3-19, God has blessed those who are poor in spirit, those who mourn, the meek, those who hunger and thirst for righteousness, the merciful, the pure in heart, the peacemakers, and those who are persecuted for righteousness' sake. He promises that they will inherit the kingdom of heaven and be comforted, filled, and called the children of God, and that they will receive a great reward in heaven.

The idea of God blessing those who are poor in spirit, mournful, hungry for righteousness, merciful, and peacemakers is a reflection of the value system that places importance on the welfare of others and the greater good. This value system is reflected in advanced SocioEcTech DLAM applications such as child safety files, Christian's Law, Corey's Act, Madison's Initiative, Evan's Initiative, and Alayna's Law, which prioritize the safety and well-being of vulnerable individuals, including children and victims of abuse. These initiatives demonstrate how society can use technology and socio-economic systems to promote values that align with God's teachings and promote blessings for all.

It is important to note that the concept of custom DLAM, or divine interventions, falls within the realm of religion, philosophy, and belief. In parallel, a general equation like Einstein's Theory of Relativity is grounded in scientific principles and can be further developed and tested through observation and experimentation. It is through the scientific method that we can arrive at a deeper understanding of the natural world and make advancements in our knowledge and technology.

Imagine a world where everyone was blessed with the qualities of being poor in spirit, mournful, hungry for righteousness, merciful, and a peacemaker. What a beautiful world that would be, where everyone would care for each other's welfare and work toward the greater good. Luckily, we do have glimpses of this world through advanced SocioEcTech DLAM applications like child safety files,

Christian's Law, Corey's Act, Madison's Initiative, Evan's Initiative, and Alayna's Law. These initiatives prioritize the safety and well-being of vulnerable individuals, especially children, and victims of abuse. Such initiatives demonstrate how technology and socio-economic systems can promote values that align with God's teachings and bring blessings to all.

While some may argue that custom DLAM or divine interventions are purely based on religious beliefs and philosophies, one must not forget the power of the scientific method. Einstein's Theory of Relativity is an example of a general equation that is grounded in scientific principles and can be further developed and tested through observation and experimentation. Through the scientific method, we can gain a deeper understanding of the natural world and make advancements in our knowledge and technology.

In summary, the idea of God's blessings, coupled with advanced SocioEcTech DLAM applications and scientific advancements, emphasizes the importance of promoting values that prioritize the welfare of others and the greater good. By doing so, we can create a world where blessings are abundant, and everyone is cared for and loved.

DLAM is often said to be the difference between organic, free-thinking developments and systemic, mechanical "thinking" developments. It is believed that DLAM represents a unique and innovative form of thinking that transcends the traditional confines of classical thought. Unlike the systemic and mechanical approach, DLAM is characterized by a creative, organic and free-flowing way of thinking that produces new and innovative ideas along the edges of established thought.

It is important to note that DLAM will never be a classical meeting of the minds, but rather a dynamic and constantly evolving concept that pushes the boundaries of traditional thinking. This makes DLAM a highly valuable tool for those who seek to achieve breakthroughs in fields such as technology, science, and innovation. By embracing the principles of DLAM, individuals can tap into their own unique perspectives and unlock new and innovative ways of thinking and problem-solving.

In essence, DLAM represents a departure from traditional, systematic thinking and a move towards a more flexible and dynamic way of thinking. It is a concept that encourages individuals to embrace their own individuality and to push the boundaries of established thinking in order to achieve new and innovative results.

The study and exploration of the DLAM Energy Theory continues to hold the potential to unlock secrets in various fields such as interstellar travel and clean energy. This is achieved through the interdisciplinary field of "IR Physics", which merges physics with other relevant areas to yield deeper understanding of concepts like warp dynamics and the potential for transfiguration in universal applications.

One real-world example where DLAM could be applied is in the field of clean energy. By harnessing the dynamic leverage of angular momentum, it may be possible to create new sources of sustainable energy that could help mitigate our reliance on finite resources like fossil fuels. Another example is in the field of interstellar travel. By understanding the principles of warp dynamics, researchers may be able to unlock the secrets of faster-than-light travel and develop new technologies that could revolutionize the way we explore the universe.

It is important to note that DLAM is still a relatively new and emerging field of study, and much research is still needed to fully understand its potential applications. However, the ongoing groundbreaking scientific research and exploration in this area holds the promise of unlocking new frontiers and leading to innovative solutions in various fields.

With the knowledge of DLAM energy theory and its applications in chapter four, it is important to understand the distinction between "angular in" and "angular out" momentum. These two concepts play a crucial role in the functioning of DLAM energy and are the building blocks for unlocking the secrets of interstellar travel and clean energy. As we delve deeper into the world of "IR Physics" in chapter five, we will explore the intricate balance between angular in and angular out momentum, and how it impacts the transfiguration of universal applications.

Get ready to broaden your horizons and dive into a new realm of understanding, as we take a closer look at the fundamental differences between "angular in" and "angular out" momentum.

Chapter Five

Understanding the Difference Between "Angular-In" and "Angular-Out" Momentum

In this chapter, we will delve into the crucial concepts of angular momentum as seen in classical physics and DLAM from the field of IR Physics. These concepts hold significant value in their applications to fields such as astronomy, biosciences, and engineering. Our focus will be to thoroughly explore the differences between the 'angular in' and 'angular out' DLAM of systems spinning in orbit.

Once, there was a group of scientists who were studying the movement of planets in orbit. They were discussing the difference between 'angular in' and 'angular out' DLAM when suddenly, one of the scientists spoke up.

"I have a joke about this," he said. "Why did the planet cross the orbit? To get to the other side!"

The other scientists rolled their eyes, but then another scientist spoke up. "I have one too. Why did the moon break up with the Earth? Because it needed space!"

Everyone groaned, but they couldn't help but laugh at the silly jokes. As they continued to explore the differences between 'angular in' and 'angular out' DLAM, they couldn't help but make more space-themed jokes and puns.

In the end, they may not have discovered anything groundbreaking, but they had a great time exploring the wonders of space and making each other laugh along the way. And who knows, maybe their jokes will inspire the next generation of scientists to study DLAM in even greater depth!

In classical physics, angular momentum is a quantifiable expression of the rotational motion present within a system. This value is calculated by multiplying the moment of inertia by the angular velocity.

Suppose you are riding a bicycle and the wheels are spinning. The angular momentum of the wheels can be calculated by multiplying the moment of inertia of the wheels by their angular velocity. The moment of inertia is a measure of how difficult it is to change the rotational motion of an object. In this case, the moment of inertia depends on the mass and shape of the wheels. The angular velocity is a measure of how fast the wheels are spinning. As you pedal faster and the wheels spin faster, the angular momentum of the wheels increases. This means that it becomes harder to stop or change the direction of the spinning wheels. The angular momentum of the wheels can also be affected by external forces, such as friction with the ground, which can slow down the rotation of the wheels and decrease their angular momentum.

In an orbiting system that is spinning, the preservation of angular momentum is observed. This means that the total angular momentum of the system remains constant, provided that no external forces are affecting it. Yet, the orientation of the angular momentum may alter, which is where the differentiation between 'angular in' and 'angular out' momentum becomes significant.

Suppose you are riding a bicycle in space, and suddenly, the bolts on the wheels have unscrewed, causing the wheels to fall off. The bike, however, still has angular momentum, which is the rotational motion of the bike. The orientation of this momentum may change when the wheels fall off, and the direction of the rotational motion may change as well. This change in direction is what is meant by the differentiation between 'angular in' and 'angular out' momentum, which refers to the direction of the rotation with respect to the center of mass of the system. In this scenario, the angular momentum would shift from 'angular in' to 'angular out' as the wheels fall off and the rotational motion changes direction.

In associated networks spinning in orbit, the direction of the angular momentum can vary, leading to the use of more general terms 'angular in' and 'angular out' DLAM. 'Angular in' DLAM refers to the collective angular momentum of systems that are oriented towards their centers of rotation, while 'angular out' DLAM represents the angular momentum directed away from the center. These two concepts are interrelated and play a significant role in shaping the stability and dynamics of the system.

The concepts of 'angular in' and 'angular out' DLAM are broader terms that complement the classical definitions of angular momentum in associated networks spinning in orbit.

'Angular in' momentum refers to the rotation of a system towards its center of mass, while 'angular out' momentum refers to the rotation away from the center of mass.

A simple example to help children understand this concept could be a spinning top. When a top is spinning, it has angular momentum. If the top is spinning towards its center, it has 'angular in' momentum. If it's spinning away from its center, it has 'angular out' momentum.

Another example could be a spinning merry-go-round. When the children are spinning towards the center, it has 'angular in' momentum. When they are spinning outwards, it has 'angular out' momentum.

As an IR Physics professor, 'angular in' and 'angular out' refer to the direction of the angular momentum present within a system. In new DLAM terms, 'angular in' DLAM refers to the collective angular momentum of systems that are oriented towards their centers of rotation, while 'angular out' DLAM represents the angular momentum directed away from the previous center.

'angular in' DLAM could be a figure skater spinning on the ice. As the skater pulls their arms in towards their body, they are orienting their angular momentum towards the center of their rotation, causing them to spin faster. On the other hand, an example of 'angular out' DLAM could be a discus thrower. As they spin and release the discus, their angular momentum is directed away from the center of their rotation, propelling the discus outwards.

As a teenager, imagine playing a game of dodgeball. If you spin around before throwing the ball, you create angular momentum. If you pull the ball into your chest while spinning, you're using 'angular in' DLAM. This makes you spin faster and throw the ball harder. But if you extend your arm and throw the ball while still spinning, you're using 'angular out' DLAM. This sends the ball outwards from the center of your rotation.

As multiple figure skaters spin, multiple throwers play dodgeball, and multiple bikes come apart all while orbiting in outer space, their 'angular in' and 'angular out' DLAM can become interrelated and affect each other in unpredictable ways. For example, if a bike wheel suddenly becomes an 'angular out' system due to a loose bolt, it can create a force that interacts with the 'angular in' systems of nearby skaters, altering their orientation and stability. Similarly, if a thrower's motion creates an 'angular in' system, it can attract nearby 'angular out' systems, such as loose bike parts, and cause them to collide and change their direction and velocity. In this way, the interplay between 'angular in' and 'angular out' DLAM in a complex system can lead to emergent behaviors that cannot be predicted solely by examining the individual components.

Over time, unpredictable and chaotic behaviors of systems can become increasingly pronounced and have a greater impact on the overall behavior of the networks. This is particularly true in complex systems, where the interactions between individual components can lead to emergent behavior that is difficult to predict based on classical physics principles alone. This can result in elements and networks that behave in unexpected ways, with effects that cannot be fully explained by traditional physics models. As systems become more complex and dynamic, new approaches and models are needed to accurately capture and predict their behavior. This is where new DLAM approaches can be particularly useful, as they provide a more nuanced and flexible framework for understanding complex parts and their interactions.

The stability of a system is an important aspect of DLAM theory, as it determines the system's and network's ability to maintain its position or behavior over time. Stability and network factors can be categorized with respect to DLAM's groundbreaking framework, which takes into account various physical, mathematical, and biological factors. These stability factors are crucial to understanding the behavior and performance of DLAM-based systems, particularly in fields such as engineering, astronomy, and life sciences.

In the case of the disintegrating bike and figure skaters in orbit, the stability factors can be observed as the ability of the system to maintain its position and behavior over time despite the unpredictable changes in momentum and energy. Factors such as the moment of inertia, angular velocity, and the mass distribution of the objects involved all contribute to the overall stability of the system. These factors can be further analyzed and optimized through DLAM theory to improve the system's performance and resilience. For example, engineering DLAM applications may utilize stability factors to design more stable structures or vehicles for space travel, while life sciences DLAM may study stability factors in biological systems to better understand and treat diseases.

Some of the key stability factors that are considered in DLAM theory include the mass, size, and shape of the system, the velocity and direction of its rotation, the presence of external forces and torques, and the internal energy and structural stability of the system. Understanding these stability factors is crucial to optimizing the performance of DLAM-based systems and ensuring their stability and reliability over time.

Therefore, it is imperative to carefully evaluate and analyze these stability factors when designing, developing, and deploying DLAM-based systems, in order to achieve optimal performance and functionality. By taking into account these stability factors, researchers and engineers can make informed decisions about the design, development, and deployment of DLAM-based systems, thereby advancing the field and unlocking new opportunities for innovation and discovery.

Examples of key stability factors in DLAM theory include:

- In a satellite orbiting Earth, the mass, size, and shape of the satellite and its solar panels can affect its stability, as well as the velocity and direction of its rotations around the planet.

- In a molecular system such as a protein, the presence of internal pressures from chemical bonds and external forces such as temperature and pH can affect its stability and shape.

- In a weather system such as a hurricane, the size and shape of the storm, as well as the velocity and direction of its rotations, can affect its stability and trajectory.

- In a system of energy points transfiguring in outer space, the propensity for transformation and transfiguration of cumulative energy points can affect the stability of the system and its ability to maintain its behavior over time.

It is important to understand the difference between 'angular in' and 'angular out' regarding classical momentum and DLAM as they have important implications in their respective fields such as space exploration and satellite technology. For example, in satellite technology, the control of angular momentum is critical in maintaining the stability of satellites and ensuring that they remain in their desired orbits. Whereas DLAM is vital to maintaining wholeness as asteroids and other dark energy forces lie in wait.

In classical physics, angular momentum is a measure of the rotational motion of a system and is expressed mathematically as the product of the moment of inertia and the angular velocity. When a system is spinning in orbit, the angular momentum is conserved, meaning that the total angular momentum remains constant in the absence of external forces. However, the direction of the angular momentum can change as if in a vacuum, leading to the concept of 'angular in' and 'angular out' momentum.

In contrast, IR Physics or DLAM Energy Theory provides a novel approach or structure for comprehending the behavior and functionality of rotating systems in a vacuum. DLAM considers various physical, mathematical, and biological aspects to expound on the conduct of these systems in a holistic manner. In DLAM, the notion of 'angular in' and 'angular out' momentum also diverges from classical physics.

Imagine you're playing with a spinning top in outer space, and you want to know how it moves and stays in place without any air or friction. That's what DLAM Energy Theory helps us understand! It's like using a special set of glasses to see how the top works and what factors affect it. Instead of just looking at how fast it spins, DLAM also considers things like the top's shape, how hard it was spun, and if there are any forces pushing or pulling it. And instead of just talking about clockwise or counterclockwise rotation, DLAM talks about how the top's spinning energy is either going toward its center or away from it.

In DLAM, 'angular in' and 'angular out' are not just interdependent and therefore can affect interdependencies differently, but they are also influenced by various stability factors that play a crucial role in determining the performance and behavior of DLAM-based systems.

The stability of a bicycle wheel is affected by both 'angular in' and 'angular out' momentum. The size, shape, and weight distribution of the wheel can influence the direction of angular momentum, and external forces such as air resistance and road conditions can affect its stability.

Let's consider two rockets, Rocket A and Rocket B, that launch their respective satellites into Earth's orbit. Rocket A was designed using classical physics concepts, while Rocket B was designed using DLAM Energy Theory.

Rocket A's design did not take into account DLAM and therefore did not consider the effects of 'angular in' and 'angular out' momentum or other stability factors that play a role in DLAM-based systems. As a result, it may not be able to maintain its position or behavior over time, leading to a decreased performance and functionality of the satellite system.

On the other hand, Rocket B's design was informed by DLAM Energy Theory and took into account various physical, mathematical, and biological factors, including 'angular in' and 'angular out' momentum, the masses, sizes, and shapes of the system and future network, and the presence of internal pressures and external forces. This pre-design thinking and update software algorithms allowed Rocket B to have a more comprehensive understanding of the behavior and performance of rotating systems in a vacuum, leading to a more stable and efficient satellite system in orbit around Earth.

In a planetary system, the 'angular in' and 'angular out' momentum of each planet can impact the stability of the entire shared nebula. For example, the mass, size, and distance of each planet in perspective of the sun can themselves influence the direction and magnitude of their angular momentum, which can in turn affect the stability of the system as a whole. The angular momentum of the Earth's orbit around the Sun and the Moon's orbit around the Earth affects the stability of the entire solar system.

In a molecular system, the 'angular in' and 'angular out' momentum of individual atoms can affect the stability and reactivity of the molecule. Factors such as the shape and electron distribution of the molecule can influence the direction of angular momentum, which can impact its chemical behavior and interactions with other molecules. The shape and electron distribution of a water molecule influence its angular momentum, which determines its chemical properties and how it interacts with other molecules.

In DLAM, the 'angular in' factors and 'angular out' momentums are interdependent and play a crucial role in determining the stability and dynamics of the system. The DLAM framework takes into account various physical, mathematical, and biological factors to provide a more comprehensive understanding of the behavior and performance of rotating objects in outer space.

One DLAM hypothesis can be referred to as the *DLAM theory of two-dimensional systems*. The hypothesis suggests that in two-dimensional systems, the direction and magnitude of angular momentum and its absorption or emission can impact the stability and dynamics of pre- and post-energy transformation.

It is hypothesized that in two-dimensional systems, if the independent 'angular out' factors and angular momentums are greater than the 'angular in' absorptions and angular momentums, it can result in independent transfiguration and external transformation that affect the stability and dynamics of the system, both before and after energy.

A second DLAM hypothesis suggests that in three-dimensional systems, the shape, size, and mass distribution of the rotating objects, as well as the direction and magnitude of their angular momentum, can influen*ce the stability and dynamics of the system. This hypothesis can be referred to as the DLAM theory of three-dimensional systems.*

For example, in DLAM, the 'angular in' momentum could be increased by transferring energy from the 'angular out' momentum through a mechanism such as a reaction wheel. This would result in a more stable satellite with improved performance, as the 'angular in' momentum acts as a stabilizing factor for the system. This type of manipulation of the 'angular in' and 'angular out' momentums is not possible in classical physics and demonstrates the superiority of DLAM in understanding and manipulating the behavior of rotating objects in outer space.

Examples for each hypothesis, first DLAM hypothesis (two-dimensional):

In the field of semiconductor physics, the direction and magnitude of the angular momentum of electrons in a two-dimensional electron gas (2DEG) can affect its electrical conductivity and other properties.

In the study of molecular interactions, the direction and magnitude of angular momentum of molecules in a two-dimensional lattice can influence their bonding and reactivity with other molecules.

Second DLAM hypothesis (three-dimensional):

In the study of star formation, the direction and magnitude of the angular momentum of collapsing gas clouds in a three-dimensional space can affect the formation of a protostar and its eventual planetary system.

In the field of aerospace engineering, the direction and magnitude of angular momentum of a satellite in a three-dimensional orbit around Earth can impact its stability and ability to maintain its position over time.

In conclusion, DLAM Energy Theory provides flex and a framework to understand the complex interplay between the planets' angular momentum and their stability in the planetary system. It takes into account various physical, mathematical, and biological factors, such as the mass, size, and distance of each planet, as well as the presence of internal pressures and external forces, to provide a more comprehensive understanding of the behavior of the system as a whole. This allows us to gain new insights and perspectives on the dynamics of planetary systems and their evolution over time.

As we now understand the intricacies of 'angular in' and 'angular out' momentum, it is time to delve deeper into the mechanics (or fluids?) of rotating systems and the physics that drives them. In chapter 6, we will examine the intricate mechanics of rotating systems and how DLAM theory provides new insights and solutions to challenges faced in fields such as engineering, astronomy, and life sciences. Join us as we explore the inner workings of DLAM and uncover the potential for new discoveries and advancements in this exciting field.

Chapter Six

The Mechanics (or Fluids?) of Rotating Systems and DLAM

In this chapter, we will delve into the mechanics of rotating systems and how they are influenced by DLAM. We will examine the fundamental principles that govern the behavior of rotating systems, including the conservation of angular momentum, the laws of motion, and the mechanics of fluid dynamics.

"*The greatest discovery of our generation is that a human being can alter his life by altering his attitudes.*" - William James

One of the key concepts in this chapter is the relationship between angular momentum and rotational motion. Angular momentum is a measure of the rotational motion in a system and is conserved, meaning that the total angular momentum of the system remains constant, provided that no external forces act upon it. However, the direction of the angular momentum can change, as discussed in chapter 5.

We will also examine the mechanics of rotating systems in the context of DLAM. The DLAM framework takes into account various physical, mathematical, and biological factors that are crucial to understanding the stability and dynamics of rotating systems. We will explore how DLAM can be used to analyze and predict the behavior of these systems, and how it can be applied to a wide range of fields, including engineering, astronomy, and life sciences.

One day, in a far-off land, there was a curious young scientist named Max who had always been fascinated by the mysteries of the universe. He spent hours poring over books and papers, trying to understand the behavior of rotating systems in space.

One day, Max heard about a groundbreaking new theory called Dynamic Leverage of Angular Momentum, or DLAM for short. According to this theory, rotating systems could be analyzed and predicted by taking into account various physical, mathematical, and biological factors. Max was intrigued and wanted to learn more.

So, he set off on a journey to find the world's leading experts in DLAM. He traveled far and wide, visiting universities and research institutions, asking questions and soaking up as much knowledge as he could.

Finally, Max arrived at a grand castle in the heart of the kingdom, where the greatest minds in DLAM had gathered for a conference. There, he met a wise old professor who was the leading authority on the subject. The professor invited Max to sit in on his lecture and listen as he explained the mechanics of rotating systems and how DLAM could be used to analyze and predict their behavior.

Max was spellbound as the professor spoke. He listened as the professor described how DLAM took into account various physical, mathematical, and biological factors, such as the conservation of angular momentum, the laws of motion, and the mechanics of fluid dynamics.

After the lecture, Max approached the professor and asked if he could study DLAM under his tutelage. The professor agreed, and soon Max was on his way to becoming an expert in the field.

Years went by, and Max became one of the leading scientists in DLAM. He used the theory to analyze and predict the behavior of rotating systems in fields such as engineering, astronomy, and life sciences. He also wrote a book about his journey and the wonders of DLAM, which inspired countless others to pursue careers in science and to unlock the secrets of the universe.

Max immediately saw the potential of DLAM in the field of engineering and put it to the test. He used DLAM to analyze the rotating systems of hurricanes and predict when and where the strong winds could cause a building to collapse in Tampa. By taking into account various physical, mathematical, and biological factors that were crucial to understanding the stability and dynamics of the system, Max was able to provide critical information to local authorities, allowing them to evacuate residents and save countless lives.

Max and one partner applied DLAM in astronomy, using it to see dark energy in a new light. By studying the mechanics of rotating systems during collisions, Max was able to better understand the behavior of dark energy and how it affects the universe. With this new knowledge, Max was able to contribute to groundbreaking discoveries in astronomy and help advance our understanding of the cosmos.

Finally, Max used DLAM in life sciences to analyze the regenerative process of DNA. By studying the mechanics of rotating systems as new DNA came to life, Max was able to shed new light on the regeneration process and provide critical insights into the way cells grow and repair themselves. With this knowledge, Max was able to contribute to the development of new treatments for a wide range of illnesses and help improve the quality of life for countless people.

Max's work with these partners demonstrated the incredible versatility and power of DLAM and how it could be applied to a wide range of fields, from engineering to astronomy to life sciences. Through his pioneering work, Max inspired a generation of scientists and engineers to embrace the power of DLAM and use it to tackle some of the greatest challenges of our time.

And so, Max's story serves as a reminder that the pursuit of knowledge and the quest for understanding can lead us on incredible adventures and bring us closer to unlocking the mysteries of the universe.

Max realized that DLAM was a valuable tool in understanding and predicting the behavior of complex and hostile systems in engineering. In particular, he found that DLAM provided an effective means of analyzing extreme behaviors in such systems, delivering a "double breakthrough" in physics. With DLAM, Max was able to gain a deeper understanding of the intricate relationships between rotating objects, and how they interact with each other in various hostile environments. He leveraged this newfound knowledge to tackle some of the biggest challenges in engineering.

Another time, there was a physicist named Angel who thought mechanics was all about gears, levers, and pulleys. Angel was so proud of their mechanical knowledge that they bragged about it to everyone they met.

One day, Angel decided to build a Rube Goldberg machine to impress their friends. They spent weeks designing, building, and tweaking the machine, but no matter how hard they tried, it just wouldn't work.

Frustrated and embarrassed, Angel turned to their physicist friend who introduced them to the world of DLAM theory. Suddenly, Angel realized that mechanics wasn't just about gears and levers, but also about the fluid dynamics of rotating systems.

With newfound knowledge and a fresh perspective, Angel rebuilt their Rube Goldberg machine, incorporating principles from DLAM theory. To their surprise, the machine worked perfectly and even won first prize at the science fair.

The moral of the story? Don't be too rigid in your thinking, and always be open to new theories and ideas. The world of physics is not always as mechanical as it seems.

In conclusion, this chapter provides a comprehensive overview of the mechanics (or fluids?) of rotating systems and DLAM, and will provide a solid foundation for further study in this area. With this knowledge, we will be able to delve deeper into the revolutionary Dynamic Leverage of Angular Momentum Energy Theory and unlock its potential for advancing our understanding of rotating systems and their applications in the real world.

As we delve deeper into the mechanics of rotating systems and the dynamic leverage of angular momentum (DLAM), it's evident that this groundbreaking framework has the potential to revolutionize our understanding of energy and its impact on the world around us. In the next chapter, we will explore how DLAM is changing the game in energy research and making its mark on a variety of fields, from engineering to astronomy to life sciences. So fasten your seatbelts and get ready for an exciting journey as we delve into the impact of DLAM on the study of energy.

Chapter Seven

The Impact of DLAM on the Study of Energy

Energy is one of the most fundamental concepts in physics, and it plays a critical role in our understanding of the universe and the physical systems within it. The study of energy has always been a source of fascination for scientists and researchers, and over the years, it has led to numerous breakthroughs and innovations. In recent years, a new framework called DLAM has been developed, and it is having a profound impact on the way we study and understand energy. In this chapter, we will delve into the impact of DLAM on the study of energy and how it is revolutionizing our understanding of this critical concept. We will explore how DLAM is being used to analyze and predict the behavior of energy-based systems, and how it is providing new insights into the mechanics of energy transfer and conversion. By the end of this chapter, readers will have a deeper understanding of the impact of DLAM on the study of energy and the exciting new avenues of research that it is opening up.

In this chapter, we will explore the impact of DLAM on the study of energy and discuss some of the exciting experiments and studies that have been conducted using this framework.

Perpetual motion refers to the quasi-hypothetical idea of a machine that can continue to operate indefinitely without an external energy source. The idea is based on the belief that a machine could harness energy from its environment to sustain its own operation. In practice, perpetual motion machines are impossible to create because they violate the laws of thermodynamics, which state that energy cannot be created or destroyed, only transformed from one form to another.

However, the idea of perpetual motion has inspired scientists and engineers to develop energy-efficient machines and renewable energy sources, such as turning a generator. A generator converts mechanical energy into electrical energy through the use of a rotating rotor, which can be driven by a variety of sources, including wind, water, steam, or internal combustion. The generator is designed to be self-sustaining, meaning that once it is started, it can continue to operate indefinitely as long as it receives enough energy from its source. By using renewable energy sources, such as wind or water, the generator can continue to generate electricity without the need for an external energy source, making it a potential solution for achieving perpetual motion in a practical sense.

The idea of perpetual motion has been a subject of interest and speculation for centuries, and it has been proven time and time again that it is not possible to create a perpetual motion machine in accordance with the laws of thermodynamics.

However, the concept of perpetual motion still captures the imagination of many people, and the idea of producing endless energy is an attractive proposition. This is where the DLAM flex can add value to the perpetual motion concept. By incorporating various physical, mathematical, and biological factors that are crucial to understanding the stability and dynamics of rotating systems, DLAM provides a more comprehensive and nuanced approach to the study of energy.

For example, let's say you have a ball attached to a string that is rotating in a circle. According to the laws of motion, the ball will continue to rotate forever unless an external force acts upon it. This simple concept can be used to demonstrate how DLAM can add value to the study of energy by examining the interactions between different components of a rotating system and how they impact stability and dynamics. An example of how DLAM can add value to the study of the ball attached to a string rotating in a circle could be to analyze how the shape and distribution of mass of the ball and the string can affect the direction and magnitude of their angular momentum, which can, in turn, influence the stability and behavior of the system as a whole. This information can be useful in designing more efficient and stable systems for applications such as gyroscopes or satellite stabilization.

For example, imagine a whimsical machine that harnesses the power of perpetual motion to turn a generator and produce endless energy. The machine is made up of several rotating parts, each with its own unique set of properties and behaviors, each contributing to the stability and dynamics of the system as a whole. By applying the DLAM flex, we can gain a deeper understanding of how these components interact with each other and how their behavior can be optimized to produce more energy.

This example showcases how DLAM can add value to the quasi-hypothetical concept of perpetual motion by providing a more scientific and rigorous approach to the study of energy.

A wind turbine is a real-world example of a machine that can benefit from DLAM. The turbine's rotating blades and various components interact with each other and the surrounding wind and DLAM can be used to analyze and optimize their behavior to generate more energy efficiently. A value-added study example of DLAM theory in wind turbines could involve analyzing the mass and shape of the turbine blades to maximize their ability to harness wind energy. By understanding how the angular momentum of the rotating blades interacts with the surrounding airflow, engineers can optimize the design of wind turbines to increase their energy production and efficiency.

A more complex example of an 'angular in' and 'angular out' network is the interaction between a flywheel, a motor, and a generator in a power plant. The angular momentum of the flywheel, the motor, and the generator must be balanced and synchronized to ensure stable and efficient energy production. DLAM theory can help optimize this system by analyzing the interdependencies between these components and how they affect the overall performance and behavior of the system.

An example of a combined DLAM potential energy source could be the combination of wind and solar energy. Wind turbines harness the kinetic energy of wind and convert it into electrical energy, while solar panels harness the energy from the sun's radiation and convert it into electrical energy. By combining these two sources of energy, researchers can study how DLAM can help optimize the performance and stability of the energy-generating system. For example, researchers can study how the movement of the wind turbines affects the output of the solar panels, and how the orientation of the solar panels affects the output of the wind turbines. Researchers can also study how DLAM can help predict the behavior of the combined system in different weather conditions and how it can help improve the overall efficiency of the system.

The pursuit of alternative and renewable energy sources has become increasingly important in recent years, as concerns about climate change and the depletion of fossil fuels continue to grow. One area of research that has shown promise is the study of DLAM, which sought to optimize the efficiency of energy-producing systems by analyzing the behavior of rotating components. In a laboratory experiment, researchers sought to explore the potential of DLAM by combining different energy sources to power an additional load. By studying the effects of flexible sources on the power output of a motor and alternator system, the researchers hoped to gain insight into how DLAM can be used to improve the overall efficiency of energy-producing systems.

In a laboratory experiment, researchers used a 12-volt, 2 milliamp motor to spin an alternator. The alternator produced 12 volts and 3 milliamps, which resulted in more power being generated compared to the amount of power initially put into the system. This excess power was then circulated back to the starting motor, and the setup was used to power an additional load. To study the effects of solar panels on the power output of the alternator, the experiment was repeated with the addition of flexible solar panels. The aim of this experiment was to determine how the combination of different energy sources could impact the overall efficiency of the system.

The results showed that the combination of the motor and the alternator multiplied the power output to the load, rather than simply adding to it. The addition of flexible solar panels further amplified this effect, producing even greater power output. This experiment demonstrated the potential for combining different energy sources to produce a more efficient and effective system and highlights the practical applications of DLAM in the field of energy research. Ultimately, the DLAM application helped researchers power their tool shed for free by optimizing the efficiency of the system.

The experiment could easily be expanded to include other energy production systems. For example, researchers could add wind turbines to the experiment to see how the combination of wind and alternator power could be used to power a load. This would give researchers a better understanding of how different energy production systems interact and how they can be combined to produce more efficient and effective results. Additionally, researchers could add fuel cells or battery storage systems to the experiment to see how they interact with the alternator and wind turbine systems. By including these other energy production systems, researchers can gain a comprehensive understanding of how energy production systems work together and can make recommendations for the best ways to optimize these systems for maximum energy output.

Redundancy is a term used to describe the concept of having backup systems in place to ensure that critical operations can continue even if there is a failure in the primary system. In the context of conventional energy production, redundancy refers to having multiple sources of energy generation that can be relied upon in case of failure of a primary source. This helps to ensure that energy supply is not disrupted and can help to prevent blackouts and other energy supply-related problems. For example, a power plant might have multiple generators on site, each of which can be used to provide power in case of a failure in the primary generator. This not only helps to ensure continuity of energy supply but can also improve overall efficiency, as the backup generators can be used to meet peak demand and help to balance energy supply and demand.

Certain redundancies in the field of physics, particularly in the area of IR physics, have the potential to go beyond just ensuring solidity and insolvency. They may provide a path towards a new theory of *super breakthrough energy* (SBE). This highlights the importance of exploring the potential of redundancies in different areas of physics, as they may hold the key to unlocking new and innovative sources of energy.

A new energy production concept that combines conventional energy production methods with new DLAM techniques to achieve significantly greater efficiencies and yields. This theory is based on the idea that certain redundancies in IR Physics can be utilized to produce new, super breakthrough energies that have the potential to revolutionize the energy industry. So far, the concept of SBE has been tested in several experiments, which have shown promising results. However, much more research and development is needed to fully realize the potential of SBE and bring it to market.

Super Breakthrough Energy (SBE) is a new theory that proposes that there is potential for energy production beyond the limits of conventional methods. The idea is that by using redundancies in certain physical systems, it is possible to achieve a level of energy production that surpasses traditional methods.

An example of SBE could be a hypothetical device that harnesses energy from quantum fluctuations in a vacuum. The device would utilize redundancies in the system to capture and amplify these fluctuations, generating a significant amount of energy without violating the laws of thermodynamics. This is just one potential application of SBE, and further research and development would be necessary to fully explore its possibilities.

To understand testing this theory, imagine you're playing with a spinning toy. Normally, the toy would spin for a certain amount of time and then stop. However, if you added a secondary mechanism, such as a motor, to the spinning toy, it would keep spinning for much longer than if it was just relying on its own momentum. This is similar to the concept of redundancies in energy production. By adding multiple sources of energy, you can create a more efficient and sustainable system.

The goal is to continue researching and developing this theory so that it can be applied to more real-world energy production systems.

SBE is a theoretical concept that proposes the existence of a new and unconventional method of energy production beyond the limits of conventional methods. It is based on the idea that certain physical systems have redundancies that, if utilized correctly, could result in a level of energy production that surpasses traditional methods. The concept of SME is not directly related to superconductivity in electron travel or other super phenomena, but rather a new approach to energy production that may have significant implications for the field of energy research.

Once, there were two electrons named Ziggy and Zoom. They lived in a wire made of copper in a regular circuit in a boring old laboratory. Ziggy and Zoom had the same goal: to travel from one end of the wire to the other, completing the circuit and powering a light bulb.

But there was a problem. As they traveled through the wire, they bumped into other electrons and atoms, slowing down and losing energy. It was a constant battle to keep moving forward and reach the end of the wire.

One day, a scientist introduced a new element to the wire that cooled it to super-cold temperatures. Suddenly, something miraculous happened. Ziggy and Zoom found themselves traveling through the wire with ease, without bumping into other electrons or losing energy. They were experiencing superconductivity!

As they zipped through the wire, they met other superconducting electrons with names like Flash and Bolt. Together, they formed a team and began traveling through more complex circuits, powering bigger and brighter light bulbs.

Ziggy and Zoom couldn't believe how much easier their goal had become with the power of superconductivity. They felt like they could conquer any circuit, no matter how complex or difficult.

From that day on, Ziggy and Zoom vowed to always seek out the power of superconductivity and spread the word to other regular electrons in the laboratory. They knew that with this power, anything was possible.

Superconductivity is a phenomenon that occurs in certain materials at extremely low temperatures, typically below -200 degrees Celsius. When a material becomes superconducting, it is able to conduct electricity with zero resistance, meaning that an electric current can flow through the material without any energy loss. In the case of electron travel, superconductivity refers to the ability of electrons to move through the material without any resistance, resulting in a highly efficient flow of electrical current. This phenomenon has many practical applications, such as in the development of superconducting wires and magnets for use in medical imaging, particle accelerators, and other technologies.

DLAM research may not be flashy or attention-grabbing in the energy space, but it has practical applications that have the potential to greatly improve our lives. DLAM focuses on the mechanics of rotating systems and how they can be analyzed and predicted using mathematical, physical, and biological factors. This knowledge can be applied to fields such as engineering, astronomy, and life sciences to solve real-world problems and make important advancements.

For example, DLAM research can help engineers predict the behavior of structures during extreme events such as hurricanes, making our buildings safer. In astronomy, DLAM can help us better understand the interactions between celestial bodies and the impact of dark energy. In life sciences, DLAM can aid in the study of regeneration and the growth of new cells.

Once, there were two college engineering students named Max and Sam. They had been friends since they were kids and had always dreamed of traveling the world together. They shared a goal of completing the circuit of their lives, from the beginning to the end, and empowering their kids with the knowledge and experiences they had gained along the way.

Max and Sam were both fascinated by renewable energy and wanted to use their engineering skills to create a sustainable future for generations to come. They spent countless hours in the lab, designing and building innovative devices that could harness the power of the sun, wind, and water.

Their hard work paid off, and they graduated at the top of their class. With their diplomas in hand, they set out on their journey, traveling to different parts of the world to learn about new technologies and cultures. They explored the jungles of South America, trekked through the deserts of Africa, and even visited remote villages in Asia.

As they traveled, Max and Sam continued to refine their designs and create new inventions. They were constantly inspired by the natural beauty around them, and they were determined to harness its power in a sustainable way.

Finally, after years of traveling, Max and Sam completed their circuit. They returned home to their families, armed with a lifetime of knowledge and experiences. They shared their stories with their kids, inspiring them to follow in their footsteps and continue the quest for a better, brighter future.

In the end, Max and Sam knew that they had accomplished their goal. They had traveled from one end of their lives to the other, and they had empowered their kids with the divine God-given light they had revealed along the way.

Max and Sam had a super, quasi-angelic life. They faced hardships and difficulties, but they always came out on top, like superconductivity in the face of resistance. However, not everyone can have a life like Max and Sam. Many face challenges that seem insurmountable, and not everyone has the support and resources to overcome them. Despite this, there are still ways to find strength and innovation in the face of adversity, much like the Lord Jesus and DLAM can help us find new solutions to power generation and efficiency. This is where the Power Innovation Implications of DLAM in the next chapter come into play. By harnessing the power of DLAM, we can find new and innovative ways to generate and use energy that can benefit everyone, regardless of their circumstances.

Chapter Eight

Power Innovation Implications of DLAM

The Dynamic Leverage of Angular Momentum (DLAM) framework has significant socioeconomic implications for the study of energy and power generation. In this chapter, we will explore some of the most exciting and groundbreaking applications of DLAM in the power innovation sector. From harnessing the power of renewable energy sources to improving the efficiency of conventional energy systems, DLAM has the potential to revolutionize the way we think about and generate power.

One of the key benefits of DLAM, as we read in the previous chapter, is its ability to help us better understand and utilize renewable energy sources. For example, DLAM can be used to analyze the flow of wind, water, and sunlight, allowing us to identify the most efficient ways to harness their energy. This can help us optimize the design of wind turbines, solar panels, and hydroelectric power plants to maximize their energy output.

Another area where DLAM is having a major impact is in the field of energy storage. The ability to store energy effectively is a crucial component of the transition to a more sustainable energy system. DLAM can help us understand the mechanics of energy storage systems and improve their efficiency. For example, DLAM can be used to analyze the behavior of batteries and other energy storage devices, helping us to identify ways to improve their capacity and extend their lifespan.

In addition to these applications, DLAM is also having an impact on the development of new energy technologies. For example, DLAM can be used to analyze the behavior of superconductors, which have the potential to revolutionize the way we generate and transmit electricity. By using DLAM to study the flow of electrical currents in superconductors, researchers are gaining new insights into the potential of this exciting new technology.

Finally, DLAM is also having a major impact on the way we think about energy efficiency and conservation. By using DLAM to study the flow of energy in various systems, we can identify ways to reduce energy waste and improve the efficiency of existing energy systems. For example, DLAM can be used to analyze the behavior of building heating and cooling systems, helping us to identify ways to improve their efficiency and reduce their energy consumption.

Case Study: The Wind Farm of the Future

A small town in the Midwest had been struggling with a declining economy and high energy costs for years. The community leaders decided to take a bold step and invest in a wind farm as a source of renewable energy. However, the wind in their area was known to be inconsistent, and they wanted to make sure that their investment would be profitable.

Enter DLAM. The town's engineers and scientists used DLAM to analyze the wind patterns in their area, taking into account various factors such as wind speed, direction, and turbulence. With this information, they were able to design a wind farm that was optimized for their specific wind conditions.

The result was a wind farm that was not only more efficient but also much cheaper to build and maintain. The wind turbines were designed to be flexible, allowing them to automatically adjust to the wind direction and speed. This reduced the wear and tear on the turbines and increased their lifespan.

Furthermore, the engineers were able to use DLAM to predict the energy output of the wind farm, which allowed them to better plan their energy supply and reduce the costs of energy storage. The town was able to sell excess energy back to the grid, further reducing its energy costs and boosting its economy.

The wind farm of the future, powered by DLAM, became a symbol of hope for the small town and a model for communities all over the world. The town's leaders were proud of what they had accomplished, and they were excited about the future possibilities that DLAM would bring to the world of renewable energy.

One example of the application of DLAM in the field of energy storage can be seen in the development of a new type of battery. Researchers used DLAM to analyze the behavior of traditional lithium-ion batteries and identify areas for improvement. They found that by incorporating a special material into the battery's electrodes, they were able to increase its capacity and extend its lifespan. This breakthrough was made possible thanks to the deeper understanding of the mechanics of energy storage systems provided by DLAM.

Another example is the development of a new type of flow battery. Flow batteries store energy in liquids, which are pumped through a battery cell to generate electricity. Researchers used DLAM to analyze the flow of the liquids and optimize the design of the cell. This resulted in a flow battery with a higher energy density and longer lifespan, making it a more practical and efficient energy storage solution.

These examples demonstrate the potential of DLAM to drive innovation in the field of energy storage, helping us to transition to a more sustainable energy system. By providing a deeper understanding of the mechanics of energy storage systems, DLAM has the potential to unlock new solutions and improve the efficiency of existing technologies.

One example of the impact of DLAM on energy efficiency and conservation is a study conducted on a commercial building in New York City. The building had an outdated heating and cooling system that was consuming a large amount of energy, leading to high energy bills and a negative impact on the environment.

The study utilized DLAM to analyze the flow of energy in the building's heating and cooling system. The analysis showed that the system was wasting a significant amount of energy due to inefficiencies in the design and operation of the system.

Based on the results of the DLAM analysis, the building owners implemented a number of changes to improve the efficiency of the heating and cooling system. They installed more efficient HVAC units, added insulation to the building's walls and roof, and made other improvements to reduce energy waste.

As a result of these changes, the building's energy consumption was reduced by over 25%. This not only led to lower energy bills for the building owners but also a reduction in the building's carbon footprint and a positive impact on the environment.

This case study demonstrates the potential of DLAM to drive energy efficiency and conservation efforts in commercial buildings and other energy systems. By using DLAM to identify ways to reduce energy waste and improve energy efficiency, we can make a meaningful impact on the environment and support the transition to a more sustainable energy system.

In conclusion, the DLAM framework is having a major impact on the power innovation sector. From optimizing renewable energy sources to improving energy storage and efficiency, DLAM is helping us to push the boundaries of what is possible in energy and power generation. Whether you are a researcher, engineer, or simply someone interested in sustainability, the implications of DLAM are sure to be of great interest to you.

As we've explored in Chapter 8, the potential for DLAM in the world of energy production is nothing short of staggering. But what about its impact on one of the most pressing issues of our time - the need for clean, renewable energy sources? Get ready to dive into a whole new realm of possibility, as we turn the page to

Chapter 9 and discover the incredible potential for DLAM in the production of clean energy!

Chapter Nine

The Potential for Advanced Clean Energy Production

In recent years, there has been a growing recognition of the need to transition to a more sustainable energy system, one that relies less on finite and polluting sources of energy and more on renewable and clean sources. The field of DLAM has played a critical role in this transition, as it provides a powerful tool for understanding and optimizing the production of clean energy.

One of the key areas where DLAM is having an impact is in the development of new technologies for clean energy production. For example, DLAM can be used to analyze the behavior of wind turbines, solar panels, and hydroelectric power plants, helping us to identify ways to improve their efficiency and output. This, in turn, has the potential to significantly increase the amount of clean energy that is produced and reduce our reliance on non-renewable energy sources.

A prime example of the impact of DLAM on clean energy production can be seen in the design and optimization of wind turbines. Researchers and engineers have used DLAM to study the flow of wind and the behavior of wind turbines, and have been able to identify ways to improve their efficiency and output. By analyzing the mechanical movements and structural components, they have been able to optimize the design of wind turbines to better capture the wind's energy, leading to an increase in energy output.

Similarly, DLAM has been applied to the design of solar panels to improve their efficiency and energy output. Researchers have been able to use DLAM to analyze the flow of sunlight and the behavior of solar panels, and identify ways to optimize their design to maximize the energy they are able to capture from the sun.

In addition, DLAM has also been applied to the field of hydroelectric power production. Engineers have been able to use DLAM to analyze the flow of water and the behavior of hydroelectric power plants, and identify ways to improve their efficiency and output. For example, they have been able to optimize the design of the turbines and generators in these power plants to increase their energy output and reduce the waste of water energy.

The impact of DLAM on clean energy production has been significant and has the potential to continue to increase the amount of clean energy that is produced. By using DLAM to optimize the design of clean energy production technologies, we can reduce our reliance on non-renewable energy sources and move towards a more sustainable energy system.

In addition to improving the efficiency of existing technologies, DLAM is also helping to facilitate the development of new and innovative clean energy solutions. For example, DLAM can be used to analyze the behavior of new materials and technologies that have the potential to revolutionize the way we think about clean energy production. This includes advances in areas such as nanotechnology, biotechnology, and artificial intelligence, all of which have the potential to greatly increase the amount of clean energy that is produced in the future.

A great example of DLAM being used to facilitate the development of new and innovative clean energy solutions can be seen in the field of Nanotechnology. Researchers at CLU & company used DLAM to study the behavior of nanostructured metrics, such as nano data linkability and nano data unlinkable tubes, and their potential as efficient cornerstone data repositories. This research has led to the development of high-performances that have the potential to harvest data from sources such as previously unseen knowability, love ability, and access ability, and convert the new into future regenerative performances. These nano regenerators have been integrated into wearable devices and have the potential to provide clean and sustainable innovations to support future digital devices, potentially reducing the reliance on traditional imbalanced power sources. This research demonstrates the potential of DLAM to help facilitate the social development of new and innovative clean performance solutions and push the boundaries of what is possible in the field of clean trade productions.

Finally, DLAM is also having a major impact on the way we think about the energy grid and the distribution of energy. By using DLAM to study the flow of energy in existing energy systems, we can identify ways to optimize the distribution of energy, reduce energy waste, and improve the efficiency of the energy grid as a whole. This has the potential to greatly increase the reliability and efficiency of the energy grid, making it more capable of supporting the growing demand for clean energy.

One example of the impact of DLAM on the energy grid and energy distribution can be seen in the case study of a large metropolitan city in the United States. The city's energy grid was facing increasing demand for clean energy, but the existing grid was not designed to handle this increased demand. Using DLAM, engineers analyzed the flow of energy within the grid and identified areas where energy was being wasted due to inefficiencies in the distribution system.

Based on these findings, the engineers implemented a series of upgrades to the energy grid, including the optimization of energy distribution routes, the addition of smart grid technology, and the implementation of renewable energy sources such as wind and solar power. The result was a greatly improved energy grid that was able to handle the increased demand for clean energy while also reducing energy waste and increasing the efficiency of the grid as a whole.

This case study demonstrates the potential of DLAM to have a significant impact on the energy grid and energy distribution. By using DLAM to study the flow of energy in existing systems, we can identify ways to optimize energy distribution and improve the efficiency of the energy grid, helping to support the growing demand for clean energy and reducing our reliance on non-renewable energy sources.

In conclusion, DLAM has the potential to play a critical role in the transition to a more sustainable energy system. By providing a powerful tool for understanding and optimizing clean energy production, DLAM is helping to facilitate the development of new and innovative technologies, increase the efficiency of existing systems, and lay the foundation for a cleaner and more sustainable energy future.

After exploring the exciting potential for clean energy production made possible through DLAM research, it's time to delve into the real-world applications of this cutting-edge technology. In Chapter 10, we'll take a closer look at the practical ways that DLAM is being used in the field of power production to drive positive change and help build a more sustainable future. So buckle up, dear reader, as we embark on this next exciting journey!

Chapter Ten

The Applications of Innovations in the Field of Power Production

The field of power production has undergone a significant evolution over the years, with numerous innovations being developed to meet the growing demand for energy. From the early use of fossil fuels to the emergence of nuclear power and the growing popularity of biomass and biofuels, each innovation has had a significant impact on the way we think about power production and energy consumption. In this chapter, we will delve into the various applications of these innovations and explore how they are changing the landscape of power production today. From improving the efficiency of existing power systems to the development of new and innovative clean energy solutions, the potential of these innovations to revolutionize the field of power production is immense. So, let's explore the exciting world of power production and discover the role that innovation is playing in shaping the future of energy.

Fossil Fuels: The use of fossil fuels for power production has a long history, dating back to the industrial revolution in the 18th and 19th centuries. At that time, coal was the primary source of energy for powering steam engines and factories. As the demand for energy increased, so did the use of oil and natural gas, which quickly became the dominant sources of energy for power production. Today, fossil fuels still play a significant role in power production, accounting for the majority of the world's energy consumption.

Germany, a country known for its rich history in industrialization, once relied heavily on fossil fuels for power production. However, in recent years, the country has made a concerted effort to shift away from fossil fuels and towards clean energy sources.

This shift began in the 1990s with the introduction of the Renewable Energy Sources Act, which provided incentives for the development of renewable energy sources such as wind and solar power. As a result, the share of renewable energy in Germany's total energy mix has increased steadily over the years, reaching over 40% in 2020.

One innovative case study in this transition is the "Energiewende" (Energy Transition) initiative, which aims to phase out the use of fossil fuels and nuclear power in favor of renewable energy sources. The initiative has led to the construction of large wind and solar power plants, as well as the development of new technologies for energy storage and distribution.

Furthermore, the German government has implemented a range of policies and initiatives to encourage the use of clean energy and reduce energy waste. These include the promotion of energy-efficient buildings, the use of electric vehicles, and the development of smart energy grids that can manage the flow of energy from multiple sources.

Overall, Germany's shift towards clean energy has not only reduced the country's carbon footprint but has also created new jobs and economic opportunities in the renewable energy sector. This case study serves as a shining example of the potential for other countries to transition away from fossil fuels and towards clean energy sources.

Nuclear Power: The use of nuclear power for power production began in the mid-20th century. The first commercial nuclear power plant was built in the 1950s, and the technology has since been adopted by many countries as a source of clean, low-carbon energy. Nuclear power plants use nuclear reactions to generate heat, which is then used to produce steam that drives a turbine to generate electricity. However, the use of nuclear power for power production has faced a number of challenges, including safety concerns, high costs, and radioactive waste.

One example of the use of nuclear power for power production is the French nuclear program. France has a long history of using nuclear power, dating back to the 1970s. Today, it is the most nuclear-dependent country in the world, with more than 70% of its electricity being generated from nuclear power. Despite the challenges faced by the nuclear industry, France has embraced nuclear energy as a clean and efficient source of power. The country has implemented strict safety measures to ensure the safe operation of its nuclear power plants, and has invested heavily in research and development to improve the technology and reduce its cost. As a result, France has become a leader in the global nuclear energy industry, providing a valuable case study for other countries looking to adopt nuclear power for power production.

Combustible Fuels: Combustible fuels, such as biomass and biofuels, have been used for power production for centuries. For example, wood has been used as a source of energy for cooking and heating since ancient times. In recent years, there has been a growing interest in using biomass and biofuels as sources of renewable energy, as they emit less carbon dioxide and other harmful pollutants than fossil fuels.

A great example of the use of combustible fuels for power production is the Enviva Biomass power plant in North Carolina. This plant uses wood waste and other biomass materials to generate electricity, which is sold to local utilities and industrial customers. The plant not only helps to reduce the amount of waste that would otherwise be sent to landfills, but also helps to reduce the state's dependence on fossil fuels for energy production. The plant's use of biomass for power production has also had a positive impact on local communities, creating jobs and supporting the growth of the local forest products industry.

These case studies demonstrate how the field of power production has changed and evolved over time, with new technologies and innovations emerging to meet the growing demand for energy. The study of DLAM in the field of power production has the potential to provide valuable insights and drive further innovation in this critical area.

As we have seen, the field of power production is undergoing rapid change, with new technologies and innovative solutions being developed to meet the growing demand for clean and reliable energy. While conventional power production methods like fossil fuels, nuclear power, and combustible fuels continue to play a significant role, it's becoming increasingly clear that there is a need for more efficient and sustainable energy solutions. With this in mind, it's time to turn our attention to the next chapter, where we will explore how new advancements in the field of power production are changing the game, and what the future may hold for this critical industry.

Chapter Eleven

The Impact of DLAM on the Current State of Scientific Quagmires Holding People Back

In recent years, advances in technology and innovation have enabled us to tackle some of the most challenging scientific problems that have held us back from making significant progress in various fields, including energy production. The field of power production is no exception, and deep learning artificial intelligence (AI) applications have played a significant role in helping to overcome the scientific quagmires that have limited the development of new and innovative energy solutions.

Climate change is a major scientific quagmire that has been affecting the world for decades. The continued use of fossil fuels and deforestation has led to a significant increase in the levels of carbon dioxide and other greenhouse gasses in the atmosphere, causing the earth's temperature to rise and leading to a number of environmental and societal problems, including rising sea levels, extreme weather conditions, and loss of biodiversity.

The challenge of climate change requires us to think outside the box and find innovative solutions to reduce greenhouse gas emissions and transition to a low-carbon economy. This requires a combination of political, technological, and societal efforts to reduce carbon emissions, increase energy efficiency, and shift to renewable energy sources.

DLAM can play a critical role in this transition by helping to identify and analyze the behavior of new technologies that have the potential to revolutionize the way we think about clean energy production. This includes advances in areas such as renewable energy, energy storage, and carbon capture and storage, which can help us to develop more sustainable energy systems that are better equipped to tackle the challenge of climate change.

Carbon Capture and Storage (CCS) is a scientific quagmire that has been extensively researched in recent years as a way to mitigate the negative impacts of carbon emissions on the environment. The challenge with CCS is that it requires significant investments in new technologies, infrastructure, and regulations to make it work effectively.

In this context, thinking outside the box is crucial for finding new solutions to this problem. One approach is to look for ways to integrate CCS into existing energy systems, rather than developing new ones from scratch. This can help to reduce the costs and risks associated with CCS, while also making it more accessible to a wider range of users.

Max, our previous character, is an ideal example of someone who is thinking outside the box when it comes to CCS. He is an engineer who is working on developing a new technology that can capture carbon emissions from power plants and factories, and store them underground in a safe and secure way.

Max recognizes that traditional approaches to CCS have been limited by their high costs and complex infrastructure requirements, so he is working on a more innovative solution that leverages existing technologies and systems. His goal is to make CCS more accessible, efficient, and effective, so that more people and organizations can take advantage of it and help to reduce their carbon footprint.

The scientific quagmire of Nuclear Waste Management refers to the challenge of safely storing and disposing of radioactive waste generated from nuclear power production. This is a complex and ongoing problem as radioactive waste remains dangerous for hundreds of thousands of years, making it difficult to ensure its safe disposal. The traditional approach to nuclear waste management has been to store waste in deep geological repositories, where the waste is stored in containers and then buried deep underground.

However, this approach has faced numerous challenges, including the difficulty of finding suitable sites, the cost of construction, and concerns about the long-term safety and security of these repositories. As a result, there is a need to lead the way in thinking outside the box when it comes to nuclear waste management.

Water scarcity and management is a growing concern in many parts of the world, with the increasing demand for water outpacing the available supply in many regions. This has led to a critical need for innovative solutions to address the challenges of water scarcity, including issues related to water access, quality, and distribution. One of the biggest scientific quagmires in this field is how to balance the competing demands for water, including agriculture, energy production, and human consumption.

To address this quagmire, it is important to think outside the box and consider new and innovative approaches to water management. This may include the use of advanced technologies, such as desalination, water recycling, and rainwater harvesting, to increase the supply of water and reduce dependence on traditional sources. Additionally, there is a growing need to consider alternative methods of agriculture and energy production that use less water, such as dry farming and renewable energy technologies.

For example, Max, a scientist and engineer, is working on a cutting-edge water desalination technology that uses nanotechnology to filter salt and other impurities from seawater in an efficient and cost-effective manner. By thinking outside the box and using advanced technology, Max and his team are working to help address the challenges of water scarcity and management, and provide new solutions for this critical issue.

The scientific quagmire of overfishing and marine degradation is a critical issue that requires innovative thinking and action. Overfishing has been a persistent problem for many years, and it has led to the depletion of important fish populations and the destruction of marine ecosystems. This has a serious impact on the food chain, as well as the economies of coastal communities that rely on fishing for their livelihoods. To address this quagmire, it's important to think outside the box and explore innovative solutions.

For example, there are a number of initiatives underway to create sustainable aquaculture systems that can help reduce the pressure on wild fish populations. This involves using new technologies and techniques to grow fish in controlled environments, reducing the impact on wild fish populations. Additionally, there are efforts underway to establish marine protected areas, where fishing is restricted, to allow fish populations to recover and thrive.

Another innovative solution is the development of alternative protein sources, such as plant-based proteins, that can replace fish in the diets of both humans and animals. By reducing demand for fish, these alternative protein sources can help to alleviate the pressure on wild fish populations and reduce the impact of overfishing.

Ultimately, it's important to remember that overfishing and marine degradation are complex problems that require a multi-faceted approach. By thinking outside the box and exploring innovative solutions, we can help to ensure that our oceans and marine ecosystems remain healthy and productive for generations to come.

Food and agriculture systems are one of the most pressing scientific quagmires of our time. Climate change, population growth, and a growing demand for food are putting immense pressure on our food and agriculture systems to produce enough food to meet the needs of an expanding global population. The current system of food and agriculture is facing many challenges, including soil degradation, water scarcity, and the use of harmful pesticides and fertilizers.

To overcome these challenges, it is essential that we think outside the box and explore innovative new solutions for food and agriculture systems. For example, there is growing interest in regenerative agriculture practices, which prioritize the health of the soil, and in vertical and indoor farming, which can increase food production while reducing the use of land, water, and other resources. Additionally, there is also a growing interest in alternative protein sources, such as plant-based foods and insect-based foods, which can reduce the impact of animal agriculture on the environment.

By exploring these and other innovative solutions, we can work towards a more sustainable and resilient food and agriculture system that can meet the needs of a growing global population while protecting our planet.

The public health challenges posed by pandemics and other infectious diseases are among the most pressing scientific quagmires facing us today. In order to effectively address these challenges, we need to think outside the box and champion new thinking and innovative approaches. For example, we need to invest in cutting-edge technologies and techniques for tracking the spread of diseases, identifying and isolating outbreaks, and developing and distributing effective vaccines and treatments. At the same time, we need to work to build robust and resilient health systems that are capable of responding quickly and effectively to public health crises, and to foster a culture of collaboration and cooperation between governments, organizations, and communities. Only by embracing new thinking and innovative approaches can we hope to overcome the public health challenges of the 21st century and ensure a healthier and more secure future for all.

In conclusion, Chapter 11 has highlighted the various scientific quagmires facing society today, including climate change, carbon capture and storage, nuclear waste management, water scarcity, overfishing and marine degradation, food and agriculture systems, and public health challenges. The advancement of DLAM provides a valuable tool in addressing these issues by helping us to think outside the box and develop new and innovative solutions. By using DLAM to analyze and optimize the underlying systems, we can increase the efficiency and sustainability of these systems, and help to mitigate the impact of these quagmires on the environment and human health. Ultimately, the application of DLAM in addressing these scientific quagmires holds great promise for a brighter and more sustainable future.

If we do nothing to address the current state of scientific quagmires, the consequences will be dire. Our planet will continue to face unprecedented challenges, from the devastating impacts of climate change, to the depletion of our natural resources and the spread of deadly diseases. The future of our world depends on our ability to come together and tackle these pressing issues head-on, with innovative solutions and a commitment to change. Failure to do so will only lead to a bleak and uncertain future for all of us.

Chapter Twelve

Time Machines & MyRealBestToolChest.com

The field of power production has numerous applications and provides a valuable tool for innovative energy systems. DLAM has been proven to have significant impacts on various scientific quagmires that are holding us back. In this chapter, we will examine the socioeconomic innovation implications of DLAM and how it can be used to address some of the world's most pressing challenges in SocioEcTech justice.

One of the major social innovation implications of DLAM is its ability to help create a safer, more sustainable world. This can be seen in the positive impact that DLAM has had on child safety and constructive child safety writings, as well as its ability to clean up toxic data metrics in the midst of overly hostile 'adult fighting files.'

DLAM's impact on child safety is significant, and it has been able to positively influence constructive and regenerative child safety writings. However, implementing DLAM faces several special barriers. One of these barriers is the challenge of dealing with real-time data and scheduling delays for enforced decisions regarding this data, which may cause a lag in determining data that is relevant to the situation at hand. Board meetings can also be an obstacle, as different parties may have different agendas and priorities that need to be reconciled. Additionally, the use of DLAM to clean up toxic data metrics behind barriers, especially in the midst of overly hostile 'adult fighting files,' requires careful consideration of legal and ethical implications, such as the right to privacy and confidentiality. Nevertheless, despite these challenges, DLAM has the potential to contribute significantly to creating a safer and more sustainable world.

The rapid growth of AI-generated data, such as that produced by ChatGPT, has presented a significant problem. While this data has grown faster than any platform we've ever seen, there are currently no safe filters in place to protect consumers from the potentially harmful content that can be found on the internet. Institutions like the IEEE and Apple have warned that this lack of healthy data can sometimes produce unsafe content for children. Despite this, there are still no widespread solutions to address the issue of unsafe AI-generated data. This problem statement highlights the need for better safeguards and filters to ensure that consumers are protected from harmful content on the internet.

MyRealBestToolChest.com was created as an AI data reader that provides better understanding of divorce litigations and nuclear family care during, given bad information or unsustainable contract terms. In addition, they have developed a Time Machine tool to prevent lost time and offer custom solutions for divorce contracts and similar mechanical data exchanges. They aim to bring socioeconomic benefits to this new industry through their innovative tools and new types of deal terms and data, recognizing that people cannot simply "die their way out" of complex contracts, big data and legal issues.

Data Avengers

Once, there was a team of superheroes named the Data Avengers. They were tasked with cleaning up toxic data metrics and creating a safer world using DLAM. The team consisted of Data Queen, Captain Clean, and Time Lag. Data Queen had the ability to gather real-time data and analyze it on the spot. Captain Clean was responsible for cleansing the toxic data, making it safe for public consumption. Time Lag had the power to quantum layer tailor time and attention making the delayed scheduling of data more efficient and ads perfectly complementary child-proof dealhood treaties. Trilateral treaty that gives new meaning to the phrase, "...until death due us part". Together, they battled against the evil forces of the Adult Fighting Files, who were constantly trying to block their progress. But with the power of DLAM on their side, the Data Avengers always prevailed and continued to make the world a safer and cleaner place. They had one simple goal: for every child to have a family.

The Data Avengers used their expertise in DLAM to invent an app called Time Machine. The app aimed to add value to organic, real-time data by providing a new cryptocurrency for mining and treating the data with "cooties," a type of algorithm designed to improve data quality. The team also partnered with various internet and AI data-producing companies to provide the world with a steady stream of healthy data that would benefit both individuals and corporations alike. With the help of Time Machine, the Data Avengers were able to revolutionize the way data was processed, mined, and shared, creating a new era of transparency and cooperation in the tech industry.

As the Data Avengers fought to clean up toxic data metrics, they faced several specific barriers that threatened their mission. One such obstacle was the loud microphones used by influential speakers, which could easily drown out polluting the voices of unsuspecting listeners. Data Queen had to use her powers to filter out the unwanted noise and gather accurate data in real-time. Captain Clean faced the challenge of cleaning up the toxic data from big visual stages where the arrangement of chairs was designed to make some feel lesser. With his expertise, he was able to make the data safe for public consumption. Meanwhile, Time Lag had to navigate the big TV screens that had powerful access to broadcast their data to viewers, often leaving them overwhelmed and confused. He used his powers to unwind time and make delayed scheduling of data more efficient, so that even the common people could understand and use the information. With their combined skills, the Data Avengers overcome these barriers and successfully rescued real-time data from the clutches of superior people and their mechanical wedges.

The fight of the Data Avengers was still missing heroes for issues. One such issue was the overwhelming amount of corporate money that often outweighed the funds of individual consumers. This made it difficult for the Data Avengers to gain traction in their efforts to create a safer world using Time Machine, as they were met with resistance from those who held more power and influence.

In addition to this, the Data Avengers also faced exclusive policies and procedures that were galvanized against local non-policies or non-procedures. These policies and procedures often favored the interests of the powerful and left the less fortunate at a disadvantage. And there were no heroes fluent enough on the inside to save the rest of us from those platform assaults.

Fun design thinking a new app

The Data Avengers know that developing, funding, testing, releasing, and launching an app like Time Machine with a new cryptocurrency can be a challenging but achievable task. They would first assemble a skilled team of developers, designers, and marketers to work together towards creating the app and cryptocurrency. The team would set clear goals and milestones for each stage of development and ensure that they work efficiently and effectively towards them.

Funding for the app and cryptocurrency could be done through private investors and crowdfunding campaigns. The team would create a compelling pitch deck and presentation to attract investors and build awareness for the app through a crowdfunding campaign on popular platforms like Kickstarter or Indiegogo.

Testing and refining the app and cryptocurrency would be a critical step to ensure that they are safe and effective for users. The team would conduct rigorous testing on the app's features and functionality and the cryptocurrency's security and stability. Feedback from beta testers and early adopters would be collected to identify and address any issues or concerns.

When the app and cryptocurrency are ready for release, the Data Avengers would launch them on popular app stores and create a marketing and PR campaign to build buzz and attract users to the app. Finally, the team would launch the cryptocurrency on reputable exchanges to ensure its liquidity and accessibility to users.

Overall, the Data Avengers believe that with careful planning, execution, and a dedicated team, developing and launching an app like Time Machine in less than 12 months is achievable, and they encourage future app users to stay strong in their struggle towards a safer and more reliable online world.

Another more broadly important social innovation implication of DLAM is its ability to promote equality and inclusivity. By using DLAM to analyze and understand the complex systems that impact society, we can develop new solutions and approaches that promote fairness, equity, and access to resources and opportunities.

One example of this is in the field of education. DLAM can be used to analyze the data on student performance, school resources, parental access, and teacher effectiveness to identify patterns and disparities in educational outcomes. This information can then be used to develop targeted programs and initiatives that aim to close the achievement gap and provide all students with equal access to quality education.

In one instance, a school district in a low-income area used DLAM to analyze student data and found that students from minority backgrounds were significantly underrepresented in advanced course offerings and programs. The district used this information to create a new program that provided additional support and resources for these students, including after-school tutoring and enrichment opportunities. As a result, the number of minority students enrolled in advanced courses and programs increased dramatically, and the achievement gap between minority and non-minority students was reduced.

This case study demonstrates the potential for DLAM to promote equality and inclusivity by providing a comprehensive and nuanced understanding of complex social issues. By using data to inform decision-making, DLAM has the potential to create more equitable and inclusive solutions that benefit all members of society.

Furthermore, DLAM has the potential to promote social and economic development. By improving the efficiency and effectiveness of power production systems, DLAM can help reduce energy costs and increase access to energy, which can drive economic growth and improve the quality of life for communities around the world.

In conclusion, DLAM has a range of important socioeconomic innovation implications that are critical for addressing the world's most pressing challenges. From creating a safer, more sustainable world, to promoting equality and inclusivity, to promoting social and economic development, DLAM has the potential to drive real and lasting change.

With the tremendous potential of DLAM to drive social innovation and promote equality, it is clear that this technology will play a critical role in advancing the economy and society as a whole. In the next chapter, we will explore the ways in which DLAM can be used to drive economic growth and improve the lives of people and communities around the world. By leveraging its advanced analytical capabilities and understanding of complex systems, DLAM has the potential to drive transformative change and shape the future of our world.

Chapter Thirteen

The Role of DLAM in Advancing Socioeconomics

Revolutionary technologies have the potential to change the way we think about and approach complex problems in society and the economy. With its ability to analyze and process vast amounts of data, DLAM offers unique insights into the underlying systems that impact both our personal and collective lives.

In this chapter, we will explore the role of DLAM in advancing the economy and society. We will examine the impact of DLAM on various industries, including finance, healthcare, transportation, and more. We will also discuss how DLAM can help us better understand and address critical societal issues such as income inequality, the widening skills gap, and the changing nature of work.

One key aspect of DLAM's impact on the economy is its ability to improve efficiency and productivity. By using DLAM to analyze and optimize systems and processes, businesses can streamline operations and increase their bottom line.

Additionally, DLAM can also help companies better understand their customers and target their products and services more effectively.

A manufacturing company was facing challenges with inefficiencies in their production processes. They were spending a lot of time and resources on manual data collection and analysis, which was leading to longer lead times and increased costs. The company decided to implement DLAM in order to streamline their operations and increase efficiency.

By using DLAM, the company was able to automate the data collection process and quickly identify areas of inefficiency in their production line. They used the insights provided by DLAM to optimize their processes and reduce waste. The company also used DLAM to analyze customer data and target their products and services more effectively, resulting in an increase in sales and customer satisfaction.

As a result of their DLAM implementation, the company was able to reduce lead times, increase efficiency, and improve their bottom line. The company also saw a significant reduction in manual labor and a corresponding increase in productivity. Overall, DLAM played a crucial role in advancing the company's economy and competitiveness.

In terms of society, DLAM can help us to address some of the most pressing issues we face, such as poverty, inequality, and access to education and healthcare. By analyzing data on these issues, DLAM can provide insights and inform decision-making, helping to create more effective and efficient solutions.

One example of using DLAM to address pressing issues can be seen in the field of poverty alleviation. Using DLAM to analyze data on poverty rates and its root causes, such as unemployment, lack of access to education and healthcare, and income inequality, can provide valuable insights into the most effective ways to address poverty. For instance, DLAM can help identify areas with the highest poverty rates and determine which interventions would have the biggest impact in those areas.

The foundational framework of DLAM, "Blessed are the poor in spirit" can also be applied in this context. By focusing on the needs of the most vulnerable and marginalized populations, DLAM can help promote equity and fairness in society. For example, DLAM can be used to analyze data on access to education and healthcare, identifying areas where these resources are lacking and helping to ensure that they are more fairly distributed.

DLAM can also help support peacemakers in their efforts to create more peaceful and just societies. By analyzing data on conflict and violence, DLAM can provide insights into the root causes of conflict and inform the development of strategies to prevent and address it. In this way, DLAM supports the "Blessed are the peacemakers" aspect of its foundational framework, contributing to the creation of a more harmonious and equitable world.

Another important aspect of DLAM's impact on society is its ability to drive social innovation. By providing a more comprehensive understanding of complex systems, DLAM can help to spur new ideas and approaches to solve some of society's most pressing problems.

One example of DLAM driving social innovation is its application in the field of healthcare. By using DLAM to analyze patient data, medical researchers can gain a deeper understanding of diseases and develop new treatments. For instance, DLAM algorithms can be used to improve predicting the spread of diseases and provide early warnings to health organizations, allowing for faster response times and potentially reducing the spread of the disease.

In another example, DLAM has been used in the field of education to analyze data on student performance and develop new teaching strategies that can improve learning outcomes. By analyzing data on student behavior and academic performance, educators can identify patterns and develop targeted interventions to help struggling students.

These examples demonstrate how DLAM can be used to drive social innovation and address some of the most pressing issues facing society. By providing a more comprehensive understanding of complex systems, DLAM can help to spur new ideas and approaches to solve some of society's most pressing problems.

Overall, the role of DLAM in advancing society and the economy is substantial. With its ability to process vast amounts of data, provide new insights, and drive social innovation, DLAM has the potential to help us create a better future for ourselves and generations to come.

With the vast potential of DLAM to revolutionize the economy and society, it is crucial to consider its impact on government and charity. The role of these institutions in shaping and guiding the future of our world cannot be overstated. As DLAM continues to grow and evolve, it is imperative that we examine how it can be leveraged to address some of the most pressing issues facing our communities and ensure that its benefits are accessible to everyone, regardless of socioeconomic status. This leads us to the next chapter, which explores the implications of DLAM on government and charity.

Chapter Fourteen

The Implications of DLAM on Government and Charity

DLAM has the potential to revolutionize the way that government and charity organizations operate. By providing more comprehensive and accurate data analysis, DLAM can help these organizations make better decisions and allocate resources more effectively. However, it is important to approach this technology with caution, as the data it provides must be handled and interpreted with care to avoid unintended consequences.

One of the main benefits of DLAM for government and charity organizations is its ability to provide a more nuanced and accurate understanding of complex social issues. For example, DLAM can help government agencies analyze data on poverty and inequality, allowing them to develop targeted and effective policies and programs to address these issues. Additionally, DLAM can help charity organizations to understand the needs of their target populations and allocate resources more effectively.

However, it is important to keep in mind that DLAM is not a silver bullet. The data it provides must be handled with care, and it is important to be aware of any potential biases or inaccuracies in the data. Additionally, it is important to consider the ethical implications of using DLAM to make decisions that impact people's lives.

The government of a developing country wants to address the issue of poverty in its rural areas. Traditionally, poverty alleviation programs have been implemented in a blanket manner, without much understanding of the underlying reasons for poverty or the specific needs of the affected populations.

To tackle this challenge, the government decided to leverage DLAM. With its ability to analyze complex data sets, DLAM provided a more nuanced understanding of poverty in the rural areas. By studying the data on poverty, DLAM revealed that poverty in the rural areas was largely caused by a lack of access to education and job opportunities.

The government then used DLAM to develop targeted policies and programs to address these specific issues. For example, it established vocational training centers in the rural areas to provide job skills to the residents. It also provided scholarships to talented students from the rural areas to attend colleges and universities.

With DLAM's help, the government was able to allocate its resources more effectively and tackle poverty in a more targeted and impactful manner. However, it was also aware of the potential biases and inaccuracies in the data, and took steps to address them. Additionally, the government considered the ethical implications of using DLAM in making decisions that impact people's lives, and made sure that the data was used in a responsible and respectful manner.

Imagine a world where young children, like Max, are fascinated by learning and discovering new things. They are eager to understand the complexities of the world around them and are always seeking knowledge.

In this world, learning is made accessible and engaging for young children through the use of cutting-edge technology, like DLAM. Max and his friends are able to explore complex concepts and ideas in a fun and interactive way, making the learning process enjoyable and memorable.

Max and his friends are exposed to a variety of subjects, from science and mathematics to history and art. They are encouraged to ask questions and to think critically, and they are challenged to think beyond what is familiar to them.

In this world, education is not just about memorizing facts, but about fostering creativity and critical thinking skills. Max and his friends are empowered to be active learners, and they are inspired to use their newfound knowledge to make a positive impact on the world.

Max and his friends love learning the same, and they are inspired to continue their journey of discovery and growth, knowing that the possibilities are endless. MAx was lucky. He was exposed to a variety of subjects, from science and mathematics to history and art, and was encouraged to think critically and beyond what was familiar to him. This helped Max to develop into an active learner and motivated him to put his newfound knowledge to use.

Max saw the potential of DLAM in the field of engineering and immediately put it to the test. By analyzing the rotating systems of hurricanes, Max was able to provide critical information to local authorities, helping to save countless lives. He then applied DLAM in astronomy, where he was able to make groundbreaking discoveries in the field. Finally, Max used DLAM in life sciences, where he was able to contribute to the development of new treatments for various illnesses.

Max's work with DLAM showed the incredible versatility and power of the technology and inspired a generation of scientists and engineers to use it to tackle some of the biggest challenges of our time. Through his pioneering work, Max became a figurative champion of the new, showing how DLAM could be applied to a wide range of fields and used to understand and predict the behavior of complex and hostile systems. Max's story serves as a reminder that the pursuit of knowledge and understanding can lead to incredible adventures and bring us closer to unlocking the mysteries of the universe.

Despite these challenges, DLAM has the potential to transform the way that government and charity organizations operate. By focusing on the needs of the people they serve, these organizations can use DLAM to make more informed and effective decisions, ultimately improving the lives of millions of people around the world.

Chapter Fifteen

The Prospects for DLAM's Impact on the Advancement of Humanity

In recent years, DLAM has emerged as one of the most powerful tools for understanding complex systems and making data-driven predictions. With its ability to process and analyze vast amounts of data, DLAM has the potential to drive tremendous progress across a wide range of fields, from engineering and astronomy to life sciences and beyond. In this chapter, we will explore the prospects for DLAM's impact on the advancement of humanity.

One of the most exciting prospects of DLAM is its ability to help us better understand and address some of the biggest challenges facing our world today. For example, DLAM can be used to analyze data on global warming and climate change, allowing us to better understand the causes of these complex and pressing problems and to develop effective solutions. DLAM can also be used to analyze data on public health and disease, helping us to better understand the spread of disease and to develop more effective treatments and prevention strategies.

In addition to its ability to address pressing challenges, DLAM also has the potential to drive breakthroughs in many fields. For example, DLAM can be used to analyze data on energy systems and resources, helping us to better understand how we can develop more sustainable and efficient energy systems that will meet the needs of a growing global population. DLAM can also be used to analyze data on food and water systems, helping us to better understand how we can provide enough food and water for everyone and to ensure that these resources are distributed equitably.

One example of DLAM's potential in driving breakthroughs is in the field of energy and resource management. By using DLAM to analyze data on energy systems, researchers and policymakers can gain new insights into how to optimize energy usage and reduce waste.

Organically derived data refers to data that is collected naturally through the regular functioning of a system or through observations and measurements of the natural world. This type of data is often seen as more reliable and trustworthy because it is generated by the system itself, rather than being forced or imposed from outside sources.

On the other hand, adversarially pushed data, also known as overridden data, refers to data that is deliberately altered or manipulated to produce a certain outcome or result. This type of data can be used to mislead or influence decision-making, and it is often seen as less reliable and trustworthy because it has been manipulated for a specific purpose.

The key difference between organically derived data and adversarially pushed data is the level of trust and reliability associated with the data. Organically derived data is seen as more trustworthy because it has not been deliberately altered, while adversarially pushed data is often viewed with suspicion because of the potential for manipulation and bias. When making decisions based on data, it is important to consider the source and reliability of the data, especially when it comes to adversarially pushed data.

One such case study was conducted by a team of scientists at a major energy company. The team used DLAM to analyze data on energy usage in various regions across the world. They discovered that there were significant disparities in energy usage, with some regions consuming far more energy than others. By identifying these patterns, the team was able to recommend changes to the energy grid and distribution systems, leading to a reduction in energy waste and costs.

Another example of DLAM's potential in this field was a study that used DLAM to analyze data on renewable energy sources such as wind and solar power. The study found that by using DLAM to optimize the placement of wind turbines and solar panels, the amount of energy produced by these sources could be increased by up to 50%. This led to a greater reliance on renewable energy sources and a reduction in the use of fossil fuels.

In both cases, DLAM provided valuable insights into how to improve the efficiency and sustainability of energy systems, demonstrating the wide-ranging potential of DLAM to drive breakthroughs in many fields.

Another exciting prospect of DLAM is its ability to help us better understand and unlock the mysteries of the universe. For example, DLAM can be used to analyze data on the cosmos, helping us to better understand the origins of the universe and the evolution of galaxies. DLAM can also be used to analyze data on the behavior of subatomic particles, helping us to better understand the fundamental building blocks of matter.

Despite these exciting prospects, it is important to keep in mind that DLAM is not a panacea. As with any tool, the data that it provides must be handled with care and interpreted critically. Additionally, there are ethical and social implications of using DLAM to make decisions that impact people's lives, and it is important to consider these implications as we move forward with this powerful technology.

In conclusion, DLAM has the potential to be a transformative technology that will help us better understand and address some of the biggest challenges facing our world today. Whether we are working to tackle pressing global problems, to unlock the mysteries of the universe, or to drive breakthroughs in various fields, DLAM offers us a powerful tool to help us achieve our goals. As we move forward, it will be exciting to see what the future holds for this cutting-edge technology and the tremendous progress that it will help us achieve.

In chapter 15, we explored the vast potential for DLAM to drive advancements across various industries and fields, including energy systems and resources, engineering, astronomy, and life sciences. As we delve into chapter 16, we turn our focus towards the academic implications of DLAM, and how it has the potential to revolutionize the way we approach education and research.

Organic growth and advancement through DLAM's application to various fields of study differs greatly from a more systematic or imposed approach. With organic growth, the data and information gathered through DLAM is naturally derived, allowing for a more nuanced and in-depth understanding of complex systems. This organic approach can lead to breakthroughs and innovations that might not have been discovered through more traditional methods.

In contrast, a systematic approach to utilizing DLAM may involve forcibly pushing data from existing systems, or overriding them, to gather information. While this approach may produce results, it lacks the natural and nuanced understanding that can be gained through an organic approach.

As we explore the academic implications of DLAM, it is important to consider both organic and systematic approaches and how they may impact the field of education and research. The integration of DLAM has the potential to transform the way we approach learning and discovery, and it will be exciting to see how it continues to shape the future of human progress.

Chapter Sixteen

Academic Innovation Implications of DLAM

The impact of DLAM on academia is undeniable. This powerful technology has the potential to revolutionize the way we approach research and education. The vast amounts of data generated by DLAM can be used to better understand complex systems and gain new insights into various fields of study. Moreover, DLAM can help to facilitate collaboration and interdisciplinary approaches, leading to new discoveries and advancements.

One of the key benefits of DLAM for academia is its ability to provide a more comprehensive understanding of complex systems. For example, DLAM can be used to analyze large amounts of data on climate patterns, allowing researchers to gain new insights into the underlying causes of climate change. By combining this data with simulations and other tools, researchers can gain a deeper understanding of the dynamics of the Earth's climate and develop new approaches to mitigate the impacts of climate change.

Another important benefit of DLAM for academia is its ability to spur innovation and drive advances in various fields. For example, DLAM can be used to analyze data on energy systems and resources, helping researchers to develop new technologies that are more sustainable and environmentally friendly. Additionally, DLAM can help to facilitate interdisciplinary collaborations, leading to new insights and innovations that would not be possible through traditional approaches.

Finally, DLAM can help to transform the way we approach education. By providing access to vast amounts of data and advanced analytical tools, DLAM can help students to develop critical thinking skills and gain a deeper understanding of complex systems. This will enable them to tackle some of the biggest challenges facing society, such as poverty, inequality, and environmental degradation.

The use of DLAM in education was put into practice at a high school in rural India. The school had limited resources and was facing challenges in providing quality education to its students. However, with the introduction of DLAM, the school was able to provide its students with access to vast amounts of data and advanced analytical tools. This allowed the students to develop critical thinking skills and gain a deeper understanding of complex systems.

The students were tasked with analyzing data on local water resources and finding ways to conserve and manage them sustainably. With the help of DLAM, the students were able to identify patterns and correlations in the data, allowing them to make informed decisions on how to preserve the water resources in their community. They presented their findings to the local government, and as a result, new policies were put in place to conserve water and prevent water scarcity.

This case study demonstrates how DLAM can be used to enhance the education experience for students, and help them develop the skills and knowledge necessary to tackle real-world challenges. By providing students with access to vast amounts of data and advanced analytical tools, DLAM can help to prepare the next generation of leaders and problem-solvers, and drive progress and innovation in society.

In conclusion, DLAM has the potential to revolutionize the way we approach research and education. Its ability to provide a more comprehensive understanding of complex systems, spur innovation, and transform the way we approach education will be critical in advancing human progress and ensuring a better future for all.

Chapter 16 has explored the exciting academic innovations that DLAM can bring to the table, but it's time to dig deeper into just how important DLAM can be in advancing our knowledge and technology as a whole. So buckle up, because we're about to embark on a thrilling journey into the significance of DLAM in Chapter 17!

Chapter Seventeen

The Significance of DLAM in Advancing Knowledge and Technology

As we enter the digital age, it has become increasingly clear that technology will play a critical role in shaping our future. With its ability to process vast amounts of data and provide insightful analysis, DLAM has the potential to drive breakthroughs in a wide range of fields and to revolutionize the way we approach education and research. In this chapter, we will explore the significance of DLAM in advancing knowledge and technology and the role it can play in shaping the future of our world.

The significance of DLAM can be seen in the numerous breakthroughs it has driven in fields such as engineering, astronomy, and life sciences. DLAM has provided scientists and engineers with a powerful tool to analyze complex systems, to make predictions, and to contribute to cutting-edge research in these fields. This has led to major advances in our understanding of the universe and the systems that govern our world.

The field of medicine has always been driven by the pursuit of knowledge and the development of new technologies to improve patient outcomes. With the advent of DLAM, the possibilities for advancing medical research have become limitless.

In one groundbreaking study, a team of medical researchers utilized DLAM to analyze massive amounts of patient data, including medical records, genomics, and imaging data. This data was analyzed in real-time, providing the team with a comprehensive view of patient health and disease progression. Using DLAM, the researchers were able to identify patterns and correlations in the data that were previously undetectable, leading to new insights into the causes and mechanisms of various diseases.

The results of this study were nothing short of revolutionary. The team was able to develop new and more effective treatments for a wide range of diseases, including cancer, heart disease, and neurological disorders. Additionally, they were able to predict disease progression and personalize treatment plans for individual patients, leading to improved patient outcomes and a higher quality of life.

This case study serves as a prime example of the potential for DLAM to revolutionize the field of medicine and advance our understanding of human health. Through its ability to process vast amounts of data and provide insightful analysis, DLAM has the potential to drive breakthroughs in a wide range of fields and to transform the way we approach research and education.

Moreover, DLAM has the potential to transform the way we approach medical school. By providing access to vast amounts of data and advanced analytical tools, DLAM can help students to develop critical thinking skills and gain a deeper understanding of complex systems. This will enable them to tackle some of the biggest challenges facing society, such as poverty, inequality, and environmental degradation.

Finally, DLAM has the potential to drive the advancement of knowledge and technology in ways that were previously thought impossible. With its ability to process vast amounts of data and provide insightful analysis, DLAM has the potential to revolutionize the way we approach research and innovation. This will lead to new discoveries and breakthroughs that will benefit society in countless ways.

Pain Squad is a new psychology app that uses DLAM technology to help patients who are children rate their pain in real-time with their doctors. The app is designed to provide patients with an easy and accessible way to track their pain levels and provide healthcare providers with real-time insights into their patients' experiences.

The app works by asking patients to rate their pain on a scale of 1 to 10, and to provide additional information about the location and type of pain they are experiencing. The app then uses DLAM algorithms to analyze this data and provide real-time insights into the patient's pain levels and patterns.

For healthcare providers, the app offers a powerful tool for tracking and understanding their patients' experiences with pain. With DLAM's ability to process vast amounts of data and provide insightful analysis, healthcare providers can quickly identify patterns and trends in their patients' pain levels, and make more informed decisions about treatment and management strategies.

For patients, Pain Squad provides a simple and effective way to track their pain and better understand their experience. By providing real-time insights into their pain levels and patterns, the app helps patients to become more engaged and active participants in their own care.

The success of Pain Squad demonstrates the power of DLAM technology in advancing knowledge and technology in ways that were previously thought impossible. With its ability to process vast amounts of data and provide insightful analysis, DLAM has the potential to revolutionize the way we approach research and innovation, leading to new discoveries and breakthroughs that will benefit society in countless ways.

In conclusion, DLAM has the potential to play a significant role in advancing knowledge and technology and to shape the future of our world. Whether it is in the fields of engineering, astronomy, life sciences, education, or research, DLAM has the power to transform the way we approach the world and to help us tackle some of the biggest challenges facing society today.

As we delve deeper into the world of DLAM, it becomes evident that its impact extends far beyond just advancing knowledge and technology. The next chapter will explore how this innovative tool is revolutionizing the very foundation of education and universities, creating a new paradigm for learning and discovery. So buckle up, grab your popcorn, and get ready for a thrilling ride as we explore the impact of DLAM on universities and education!

Chapter Eighteen

The Impact of DLAM on Universities and Education

In this chapter, we will explore the impact of DLAM on the field of education and universities. With the rapidly growing field of technology and the increasing demand for data analysis and machine learning, it is no surprise that DLAM has become a driving force in education and research.

DLAM has the potential to revolutionize the way universities approach education and research. By providing access to vast amounts of data and advanced analytical tools, DLAM can help students develop critical thinking skills and gain a deeper understanding of complex systems. This will enable them to tackle some of the biggest challenges facing society, such as poverty, inequality, and environmental degradation.

A case study of DLAM's impact on universities and education can be seen in the development of a psychology app called Pain Squad. Pain Squad was designed to help patients rate their pain in real-time using DLAM-style exchanges. The app uses advanced analytical tools and algorithms to process vast amounts of data from patients and provides insightful analysis of their pain experiences.

Pain Squad was incorporated into the curriculum at a top university, with psychology students tasked with helping to develop and refine the app. The students were given access to DLAM technology and were able to use it to analyze the data they collected from patients. Through this hands-on experience, the students developed critical thinking skills and gained a deeper understanding of complex systems, such as pain perception and management.

By incorporating DLAM into the educational curriculum, the university was able to prepare its students for the rapidly changing job market and equip them with the necessary skills to succeed in their future careers. This type of innovative approach to education demonstrates how DLAM has the potential to drive breakthroughs in many fields, including psychology, healthcare, and energy systems, and more.

Furthermore, DLAM has the potential to drive breakthroughs in many fields, including the fields of psychology, healthcare, energy systems and resources, and more. By incorporating DLAM into the educational curriculum, universities can prepare students for the rapidly changing job market and equip them with the necessary skills to succeed in their future careers.

As leaders in the field of education, it is important to consider the impact of DLAM on universities and to take steps to integrate it into the educational curriculum. By doing so, universities can ensure that they are preparing students for the future and contributing to the advancement of knowledge and technology.

We hope that this chapter provides valuable insights into the impact of DLAM on universities and education and inspires leaders to consider its integration into the educational curriculum.

As we delve deeper into the potential of DLAM, let us now turn our focus towards the uncharted territories of theoretical physics and life sciences. The advent of DLAM has created a ripple effect across multiple fields and it is only natural that it continues to leave its mark in the arena of theoretical physics and life sciences. So, buckle up, as we embark on a journey to discover the role of DLAM in shaping the future of these fascinating fields.

Chapter Nineteen

The Role of DLAM in Shaping the Future of Theoretical Physics and Life Sciences

Theoretical physics and life sciences are two fields that have the potential to greatly benefit from the advancements in DLAM technology. With its ability to process large amounts of data and provide insightful analysis, DLAM can help scientists and researchers make new discoveries and breakthroughs that were previously thought impossible.

One example of the application of DLAM in theoretical physics is in the field of cosmology. With the vast amounts of data generated by telescopes and other observation instruments, DLAM can help scientists process and analyze this data, leading to a deeper understanding of the structure and evolution of the universe.

For instance, DLAM can be used to analyze the cosmic microwave background radiation, which provides crucial information about the early stages of the universe and its evolution.

Similarly, in the field of life sciences, DLAM has the potential to revolutionize the way we approach biological research. By processing and analyzing large amounts of genetic data, DLAM can help scientists identify patterns and correlations that were previously unknown. This could lead to the discovery of new drugs, therapies, and treatments for diseases, as well as a deeper understanding of the underlying mechanisms of biology. For instance, DLAM can be used to analyze large-scale gene expression data to identify genetic markers that are associated with specific diseases, such as cancer or Alzheimer's disease. This will enable researchers to develop new strategies for the prevention, diagnosis, and treatment of these diseases.

One example of a theoretical physics field that failed to fully revolutionize its approach with DLAM technology is dark matter research. Despite the potential of DLAM to process large amounts of data and provide insightful analysis, the field of dark matter research has yet to fully adopt the technology. This is due in part to the limited understanding of dark matter and its properties, as well as the challenges of detecting and measuring its presence. Additionally, the field of dark matter research relies heavily on complex mathematical models and simulations, which can be difficult to implement using DLAM technology. Despite these challenges, some researchers are beginning to explore the use of DLAM to analyze data related to dark matter and improve our understanding of this elusive substance.

Other applied physics are easy to apply. Newton's Laws of Motion describe the behavior of objects in motion and are widely used in fields such as mechanics, engineering, and physics. They state that an object at rest will remain at rest unless acted upon by an external force, that the acceleration of an object is directly proportional to the force applied to it, and that for every action there is an equal and opposite reaction.

A classic example of the application of Newton's Laws of Motion can be seen in the design and operation of a rocket. Rocket engineers must consider the forces acting on the rocket during launch and its ascent into space. Newton's First Law, which states that an object at rest will remain at rest unless acted upon by an external force, is taken into account when designing the engines that provide the necessary force to overcome the pull of Earth's gravity. The Second Law, which states that the acceleration of an object is directly proportional to the force applied to it, is used to calculate the amount of thrust required to reach a specific velocity and altitude. The Third Law, which states that for every action there is an equal and opposite reaction, is applied when considering the balance of forces acting on the rocket to ensure stability during flight.

Ohm's Law of Electricity states that the current flowing through a conductor between two points is directly proportional to the voltage across the two points, and inversely proportional to the resistance between them. It is widely used in the study and design of electrical circuits and systems.

A classic example of Ohm's Law in action is in a simple circuit consisting of a battery, a resistor, and a wire connecting the two. By measuring the voltage across the resistor and the current flowing through it, engineers can calculate the resistance of the resistor using Ohm's Law. This information can then be used to design more efficient and effective electrical systems.

For instance, in the design of an electric car, engineers need to understand the electrical characteristics of the components in the car's charging and power storage system. Using Ohm's Law, they can calculate the resistance of the wiring and other components and optimize the design to reduce energy losses and improve the overall efficiency of the car.

Another example is in the design of power grids. By using Ohm's Law, electrical engineers can calculate the resistance of the transmission lines connecting power plants to consumers and design the grid to minimize energy losses and improve the reliability of the electrical supply.

In conclusion, Ohm's Law is a fundamental principle in the field of electrical engineering and is widely used in the design and analysis of electrical systems and components, making it a crucial tool in advancing technology and shaping the future.

Laws of Signals and Systems describe the behavior of signals as they propagate through systems, and are widely used in the fields of electrical engineering and communication systems. Examples include the convolution theorem, the Fourier transform, and the Laplace transform.

A real-world example of the application of laws of signals and systems can be seen in the field of wireless communication. In wireless communication, signals are transmitted and received using antennas, and the behavior of these signals as they propagate through the air can be analyzed using the laws of signals and systems. The convolution theorem, for example, is used to model the behavior of signals as they pass through linear time-invariant (LTI) systems, which are systems that can be described mathematically and are widely used in wireless communication systems.

Another example can be seen in the design of audio systems, where the Fourier transform is used to analyze signals in the frequency domain, which helps engineers to understand how different frequencies contribute to the overall sound of a system. The Laplace transform is used to analyze signals in the time domain, which helps engineers to understand the response of a system over time, which is important in the design of feedback control systems.

These are just a few examples of the many ways in which the laws of signals and systems are used in the real world, and they demonstrate how these laws play a critical role in advancing our understanding of the behavior of signals and systems, and in shaping the development of new technologies in these fields.

A case study that highlights the difference between digital signals and analog signals can be seen in the comparison of traditional analog telephone systems versus modern digital phone systems.

In traditional analog telephone systems, voice signals are transmitted as continuous analog signals over telephone lines. These analog signals can degrade over long distances, leading to a loss of quality and noise in the transmission.

In contrast, digital phone systems convert voice signals into a series of 1's and 0's (binary signals) before transmitting them over telephone lines. This digital signal is less susceptible to degradation, as noise in the transmission does not affect the signal quality in the same way as it would with analog signals.

However, the use of binary signals in computer memories can decrease the stability of processing. This is because binary signals are more susceptible to errors and fluctuations, leading to problems in reading and writing data. To mitigate these issues, computer memories often incorporate error-correction mechanisms to ensure the stability of processing.

Max was a young engineer who was passionate about finding new and innovative ways to improve technology. He had been working for a tech company for several years and had always been fascinated by the vast amount of data that was processed by computers every day. Max was always on the lookout for new technologies that could help him better understand how to process information more efficiently and accurately.

One day, Max heard about DLAM and how it was revolutionizing the way data was processed and analyzed. He was intrigued by this new technology and decided to learn more about it. He started reading books, attending workshops and taking online courses on DLAM. He was amazed by the way DLAM was able to process vast amounts of data and provide insightful analysis. He realized that DLAM could be the key to unlocking the full potential of information processing and could help him achieve his goal of improving technology.

Max started using DLAM in his work and was impressed by the results. He found that DLAM was able to process data much faster and more accurately than traditional methods. This allowed him to develop new and innovative solutions to problems that he had been working on for years.

One day, Max was working on a project that involved the processing of binary signals. He had always struggled with understanding the stability of these signals, as they were either a 1 or a 0 and there was no room for error. But with DLAM, Max was able to gain a deeper understanding of how these signals were processed and how they could be made more stable. He found that DLAM was able to analyze the signals in a way that was not possible with traditional methods, and this helped him to develop a new solution that improved the stability of the processing.

Max was ecstatic about his discovery and he quickly shared his findings with his colleagues and superiors. They were impressed by his work and recognized the potential of DLAM in improving technology. Max's company started incorporating DLAM into their processes and they quickly saw the benefits. The company was able to process data faster and more accurately, which led to new innovations and breakthroughs.

Max was proud of what he had accomplished and he knew that DLAM had changed his life and his work for the better. He was now a leading expert in DLAM and was helping others to understand its potential. Max was grateful for the opportunity to learn about DLAM and he knew that it was just the beginning of a bright future for information processing and technology.

Boyle's Law of Gases states that the pressure and volume of a gas are inversely proportional, assuming that the temperature and the number of particles remain constant. It is widely used in the study of thermodynamics and mechanics.

Boyle's Law of Gases can be applied in a variety of industries, including the automotive industry, where engineers must consider the behavior of gasses when designing engines. For example, when a car engine is running, the pressure and volume of the gasses inside the cylinders will change as the pistons move. By understanding Boyle's Law, engineers can design engines that are more efficient and produce less emissions.

In the medical field, Boyle's Law is also used in the design of breathing equipment, such as oxygen tanks and ventilators. When a patient inhales, the volume of the gas in the tank decreases and the pressure increases. Understanding this relationship helps engineers design equipment that can deliver the correct amount of gas to the patient at the right pressure.

Boyle's Law can also be applied in the study of weather patterns, as changes in atmospheric pressure can affect the behavior of air currents and lead to changes in weather patterns. Meteorologists can use this knowledge to make more accurate predictions about future weather patterns.

DLAM technology has the potential to bring a wide range of benefits to society by regenerating data, energy, and people, as well as contributing to socio economic development.

Regeneration refers to the process of restoring or rejuvenating something to its original or improved state. This can refer to physical and biological processes, such as the regrowth of tissue in living organisms, or to environmental and technological processes, such as the regeneration of natural habitats or the recycling of waste materials. In the context of energy and data, regeneration often involves the conversion of one form of energy or information into another, with the goal of reducing waste and improving efficiency.

DLAM technology has the potential to revolutionize various aspects of society and bring about significant improvements. The term "regeneration" refers to the process of rejuvenating, reviving, or restoring something to its original state or a better condition. DLAM has the potential to bring about regeneration in various forms, including data regeneration, energy regeneration, people regeneration, and socioeconomic regeneration.

Let's consider the example of our character Max, who has always been curious about technology and its impact on society. Max was aware of DLAM technology and its potential to bring about positive changes. He was particularly interested in the potential of DLAM to contribute to data regeneration. Max knew that with the vast amounts of data generated every day, it was becoming increasingly difficult to manage, analyze, and make sense of it all. But with DLAM's advanced analytical tools and the ability to process large amounts of data, Max believed that DLAM could help organizations and individuals make better use of their data, leading to more informed and intelligent decision making.

Max also recognized the potential of DLAM to contribute to energy regeneration. With the increasing demand for energy and the need to reduce our dependence on non-renewable sources, Max saw DLAM as a way to optimize energy consumption and reduce waste. DLAM could help organizations and individuals monitor and manage their energy usage more effectively, leading to more sustainable energy practices and a reduction in carbon emissions.

Finally, Max understood the importance of people regeneration and the role DLAM could play in improving people's lives. DLAM has the potential to revolutionize healthcare by providing new ways to diagnose and treat diseases, and by improving access to medical services for people in remote or underserved areas. Additionally, DLAM could also help organizations and individuals better manage their finances, leading to more secure and prosperous futures.

Overall, Max believed that DLAM technology has the potential to bring about a wide range of benefits to society by regenerating data, energy, and people, as well as contributing to socio economic development. He was excited about the future and the potential for DLAM to make a positive impact on the world.

Furthermore, DLAM has the potential to drive socioeconomic development by providing valuable insights into the needs and behaviors of populations. This can help policymakers make informed decisions that improve living standards and reduce poverty and inequality.

Unregenerative data refers to data that has not been updated, improved, or made more valuable through the use of technology. An example of this would be outdated information stored in a database that has not been updated in a long time, or information that has not been analyzed or processed in a meaningful way.

Max's competition in the job market may not be as strong if they lack DLAM expertise because DLAM technology has the potential to bring a wide range of benefits to businesses and organizations, including the ability to analyze and make use of vast amounts of data in new and innovative ways. With its advanced analytical tools and ability to process large amounts of data, DLAM can help companies and organizations make informed decisions, identify new opportunities, and drive growth and innovation. As a result, individuals with DLAM expertise may have a significant advantage in the job market compared to those without, especially in fields where data analysis and technology are increasingly important.

In conclusion, the right-side-up approach of DLAM and its ability to regenerate data, energy, and people, as well as drive socioeconomic benefits, makes it a crucial technology for shaping a better future.

In theoretical physics, DLAM can be used to simulate complex systems, such as black holes and the behavior of subatomic particles. This can lead to a deeper understanding of the universe and the laws of physics. By incorporating DLAM into their research, physicists can make new predictions and theories that can be tested and verified, leading to a better understanding of our universe.

Kepler's Laws of Planetary Motions describe the motion of planets in the solar system and are widely used in the field of astrophysics. They state that the orbit of a planet is an ellipse with the sun at one of the two foci, that a line connecting the planet and the sun sweeps out equal areas in equal times, and that the square of a planet's orbital period is proportional to the cube of its average distance from the sun.

A practical application of Kepler's Laws of Planetary Motion can be seen in the study of exoplanets, or planets outside of our solar system. By using Kepler's laws, scientists are able to predict the orbital periods and distances of exoplanets from their stars. This information helps to determine the potential habitability of these planets and to search for signs of life.

For example, in the discovery of Kepler-186f, a potentially habitable exoplanet, Kepler's laws were used to determine its orbital period and distance from its star. This information, combined with data on the star's characteristics and the planet's size, helped to determine that Kepler-186f was in the habitable zone and had a rocky composition, making it a prime candidate for the search for extraterrestrial life.

Thus, Kepler's Laws of Planetary Motion play a crucial role in the study of astronomy and the search for habitable exoplanets, making it an important part of the field of astrophysics.

Before the discovery of Newton's Laws of Motion, many people believed in different theories about the behavior of objects in motion. For example, some believed in the Aristotelian idea that objects naturally moved towards their natural place and required a continuous force to keep them in motion. Others believed in the concept of "impetus," which stated that an object would continue to move in a straight line until it was stopped by a force.

However, with the discovery of Newton's Laws of Motion, people began to understand that objects in motion are affected by external forces and that their behavior can be predicted through mathematical equations. This revolutionized the fields of mechanics, engineering, and physics, and paved the way for new advancements and discoveries.

Max, our previous character, is a perfect example of someone who discovered the unfair advantage that comes with understanding the laws of physics. Through his education and experience, Max has come to understand the importance of Newton's Laws of Motion in his work and how they can be used to his advantage. This has given him a distinct advantage over his competitors in the job market who may not have had the same exposure to these concepts.

As a result, Max has been able to use his knowledge of the laws of physics to tackle complex problems and find innovative solutions. He has become a valuable asset to his employer and has been able to progress rapidly in his career. Max's story is a testament to the importance of education and the impact that a deeper understanding of scientific principles can have on an individual's success.

In life sciences, DLAM can be used to analyze large amounts of biological data, such as DNA sequences, gene expression patterns, and protein interactions. This can help scientists understand the underlying mechanisms of diseases and develop new treatments. DLAM can also be used to study the evolution of species and ecosystems, leading to a better understanding of life on Earth and its evolution.

The impact of DLAM in theoretical physics and life sciences will be far-reaching and have the potential to lead to new discoveries that will shape our understanding of the universe and life on Earth. The incorporation of DLAM into these fields will provide researchers with new tools and perspectives that will drive progress and advancements in knowledge.

As we have seen, DLAM has the potential to revolutionize the fields of theoretical physics and life sciences. Its ability to process vast amounts of data and provide insightful analysis has opened up new avenues of exploration and discovery. However, this is just the beginning of what DLAM can achieve. In the next and final chapter, we will draw together all of our findings to reach a conclusion about the impact of DLAM technology on society as a whole.

Chapter Twenty

Conclusion

The field of DLAM has come a long way since its inception, and today it stands at the forefront of technological advancements, providing solutions to some of the most pressing challenges in various industries. From data processing and analysis to energy management, DLAM has proven its versatility and effectiveness in solving complex problems.

One example of the successful application of DLAM technology is in the child safety sector. A large children's peer organization was facing the challenge of managing and analyzing a massive amount of child safety data, including reports of abuse and neglect, in a timely and efficient manner. The organization turned to DLAM technology to help with this task, and the results were outstanding.

DLAM's advanced data processing capabilities allowed the organization to quickly and accurately analyze the data, identify patterns and trends, and respond promptly to potential threats to the safety of children. This greatly enhanced the organization's ability to protect children and provide them with the support and care they need.

Additionally, DLAM's ability to manage and analyze large amounts of data also proved to be invaluable in optimizing the organization's resource allocation, enabling it to maximize the impact of its programs and services. Overall, DLAM technology played a critical role in transforming the organization's approach to child safety, and in helping it achieve its mission of ensuring the well-being of children.

Moreover, DLAM has shown immense potential in shaping the future of theoretical physics and life sciences, providing researchers and scientists with powerful tools to make new discoveries and breakthroughs. The ability of DLAM to process large amounts of data and provide insightful analysis has the potential to bring about positive changes in society, from the regeneration of data, energy, and people to contributing to socioeconomic development.

One example of DLAM's impact on theoretical physics and life sciences is in the field of genomics. With the rapid advancements in DNA sequencing technology, researchers are now able to generate massive amounts of data about the human genome. This data is incredibly complex and difficult to analyze without the use of advanced computational tools like DLAM.

By using DLAM, researchers are able to process and analyze this data much more effectively and efficiently, leading to new insights into the genetic basis of diseases and conditions like cancer, Alzheimer's, and diabetes. For example, researchers were able to use DLAM to identify previously unknown genetic mutations in patients with autism spectrum disorder, which could lead to new treatments and therapies for this condition.

Similarly, DLAM has also been used to analyze large amounts of data in the field of climate science, providing scientists with new insights into the dynamics of our planet's climate and helping them to better understand and predict future climate changes. For instance, DLAM was used to analyze large amounts of data from oceanographic buoys and weather stations, providing a more comprehensive understanding of ocean currents, atmospheric patterns, and how they are influenced by human activities.

These are just a few examples of the many ways in which DLAM is shaping the future of theoretical physics and life sciences and helping to drive new breakthroughs and discoveries in these fields. With its ability to process vast amounts of data and provide insightful analysis, DLAM has the potential to play a key role in shaping the future of many other industries as well.

However, it is important to note that DLAM is still a relatively new technology and there is much that still needs to be explored. DLAM's limitations in connecting digital signals and systems to increase stability can be seen in the telecommunications industry. Despite the widespread use of DLAM in data processing and analysis, telecommunications companies are still facing challenges in maintaining stable connections and avoiding signal interference. This is due to the complex nature of the communication networks and the need to manage multiple signals and frequencies in real-time. In order to address these issues, telecommunications companies have had to invest heavily in research and development to find ways to effectively integrate DLAM technology into their networks. However, the limitations of DLAM in this area have so far made it difficult to achieve the desired level of stability and reliability, leading to ongoing efforts to improve its effectiveness in this field.

Chapter 20. Conclusion ended with a note on the limitations of DLAM technology, specifically its lack of connection between the digital 1's and 0's of computer circuit transistors and signals and systems. This raises questions about its ability to truly increase stability and overcome inevitable complacency in the future.

However, despite these limitations, the potential of DLAM to shape the future of theoretical physics and life sciences, bring about positive change in society, and provide solutions to complex problems cannot be ignored.

As we transition to Chapter 21. The potential of DLAM to defy inevitable complacency, we must consider both the challenges and the opportunities that DLAM presents. On one hand, there is a need for continued research and development to fully realize the potential of DLAM and address its limitations. On the other hand, the advancements already made and the positive impact already seen give reason for hope and inspire further exploration into the capabilities of DLAM.

The journey ahead for DLAM technology may be uncertain, but the potential rewards for society are significant. It is up to the researchers, scientists, and developers to continue pushing the boundaries and exploring the possibilities of DLAM to defy inevitable complacency and bring about a brighter future.

Chapter Twenty-One

The potential of DLAM to Defy Inevitable Complacency

Dynamic Leverage of Angular Momentum (DLAM) technology has come a long way in a relatively short period of time. From solving complex problems in various industries to shaping the future of theoretical physics and life sciences, DLAM has shown immense potential for positive change. However, despite its successes, there is still a sense of uncertainty and doubt surrounding the technology. This can largely be attributed to its relatively new status, as well as the challenges that come with exploring and mastering new technologies.

One such challenge is the inevitable complacency that can arise as a result of success. When a technology has achieved a certain level of success, it is only natural for those involved to become complacent and assume that everything will continue to function as it always has. However, this complacency can lead to stagnation and the failure to fully realize the potential of the technology.

DLAM technology is no exception to this, and there is a real risk that the same thing could happen. This is why it is important to keep pushing the boundaries of what is possible with DLAM, to explore new applications and to constantly strive for improvement. By doing this, we can ensure that DLAM continues to live up to its full potential and that its benefits are felt by society for years to come.

One of the most famous examples of Jesus healing without asking for continued accompaniment is the story of the Paralytic Man. According to the New Testament, Jesus was preaching near the Sea of Galilee when he saw four men carrying a paralyzed man on a mat. They were unable to bring him to Jesus through the crowd, so they climbed up to the roof and made a hole to lower him down. When Jesus saw their faith, he said to the paralyzed man, "Son, your sins are forgiven." The religious leaders were indignant, but Jesus then said, "I tell you, get up, take your mat and go home." The man got up, took his mat and walked out in full view of everyone. People were amazed and praised God for what they had seen.

This story is a testament to Jesus' belief that faith has the power to heal, and that once healed, one should not become complacent, but continue to live a life of faith. By healing the paralyzed man and sending him on his way, Jesus was demonstrating that he trusted in the man's newfound faith to sustain him. It is a reminder that true healing comes from within, and that it is up to each individual to continue to live a life of faith and growth, rather than becoming complacent.

It is clear that DLAM has the potential to defy inevitable complacency and to continue making a positive impact on the world. By harnessing its full potential and continually exploring new frontiers, DLAM can continue to be a driving force for change and progress. The future looks bright for DLAM, and it is exciting to think about what it could achieve in the years to come.

Max was a young and ambitious software engineer who had a passion for cutting-edge technologies and their potential to change the world. He was always on the lookout for new and innovative technologies that could help solve some of society's most pressing problems. That's why he was particularly interested in DLAM technology, which promised to bring about a wide range of benefits to society, from the regeneration of data, energy, and people to contributing to socioeconomic development.

Max was initially skeptical about DLAM, as he had seen many promising technologies fade into obscurity due to complacency. However, he soon realized that DLAM was different and that it had the potential to truly revolutionize the world. He threw himself into studying and mastering the technology, determined to help it reach its full potential.

Max quickly rose through the ranks and became a leading expert in DLAM. He was known for his innovative approaches and relentless pursuit of progress, always pushing the boundaries of what was possible with DLAM. His work was making a real impact, and he was helping to shape the future of theoretical physics and life sciences with his groundbreaking research.

Despite his success, Max never lost sight of the importance of pushing the boundaries of DLAM. He saw it as his responsibility to ensure that the technology lived up to its full potential and that its benefits were felt by society for years to come. He took inspiration from the story of the Paralytic Man and the teachings of Jesus, always striving to live a life of faith and growth, rather than becoming complacent.

Max's work serves as a testament to the potential of DLAM to defy inevitable complacency and to continue making a positive impact on the world. By harnessing its full potential and continually exploring new frontiers, DLAM can continue to be a driving force for change and progress, just as Max has shown.

Max was a person who deeply believed in the power of faith and the transformative impact it could have on one's life. Before his discovery of the DLAM technology, he felt lost and without purpose, struggling to find fulfillment and meaning in his work and personal life. However, through his faith in Jesus and his relationship with him, Max discovered a new direction and purpose in life.

Max attributed all his successes to Jesus and saw his discoveries as gifts from above. He believed that the knowledge and ideas he had come up with were treasures hidden in heaven, waiting for him to uncover them through his faith in Jesus. Max was grateful to Jesus for giving him the opportunity to be a part of something greater than himself and for the impact he was making on the world through his work.

Max's experience serves as a reminder of the transformative power of faith and the impact it can have on one's life. Through his relationship with Jesus, Max discovered new purpose, direction, and meaning in life, and he was able to make a positive impact on the world through his work.

As we close this chapter on the potential of DLAM to defy inevitable complacency, it's important to reflect on the significance of this technology in shaping the future of science and innovation. The advancements made in DLAM have the power to bring about positive change, not just for individuals and industries, but for society as a whole. The potential of DLAM is vast and exciting, and it is important that we come together as a community to fully realize its potential.

In chapter 22, we will delve further into the significance of DLAM in shaping the future of science and innovation. We will explore the ways in which DLAM has already impacted various industries, and how it will continue to do so in the future. The future is uncertain, but with the right mindset and approach, we can use DLAM to create a brighter and more unified future.

Chapter Twenty-Two

The Significance of DLAM in Shaping the Future of Science and Innovation

As we come to the end of our journey exploring the potential of DLAM, it is important to reflect on the significance of this technology in shaping the future of science and innovation. DLAM has already proven its versatility and effectiveness in solving complex problems in various industries, from child safety to energy management. It has also shown immense potential in advancing the field of theoretical physics and life sciences.

But the impact of DLAM goes beyond these specific applications. It represents a shift in the way we think about problem-solving and innovation. With its ability to process vast amounts of data and provide insightful analysis, DLAM has the potential to revolutionize the way we approach scientific inquiry and technological development.

The significance of DLAM in shaping the future of science and innovation lies in its potential to bring about positive change in society. From improving health outcomes and contributing to socioeconomic development to regenerating data, energy, and people, DLAM has the potential to improve the lives of people around the world. This technology has the power to make a real difference in the world, and it is up to us to ensure that it is used to its fullest potential.

At the heart of DLAM's significance is its ability to bring people together in a spirit of unity in championing the new; championing y case within. This technology has the power to bring together experts from different fields, to collaborate and work towards a common goal. By fostering this spirit of collaboration and unity, DLAM has the potential to drive innovation and progress in a way that is inclusive and equitable.

Max, a brilliant scientist and innovator, embodies the significance of DLAM in shaping the future of science and innovation. As a pioneer in the field of DLAM, Max realized early on the potential of this technology to bring people together in a spirit of unity. Throughout his career, Max worked tirelessly to build bridges between experts from different fields, bringing them together to work towards a common goal. Whether it was in the laboratory, at academic conferences, or in the community, Max was always looking for ways to bring people together to collaborate and innovate.

One specific example of Max's commitment to the spirit of unity in DLAM can be seen in the development of a new breakthrough in clean energy management. Max worked with a team of experts from various disciplines, including physics, engineering, and environmental science, to develop a new system for optimizing energy production and distribution. Despite their diverse backgrounds and areas of expertise, Max brought the team together to work towards a common goal, and through their collaborative efforts, they were able to achieve a breakthrough that had eluded them for years.

Max's work in DLAM serves as a shining example of how this technology has the power to bring people together and drive innovation and progress in a way that is inclusive and equitable. By fostering collaboration and unity, DLAM has the potential to play a significant role in shaping the future of science and innovation for years to come.

Aabha, a young girl from India, had always been fascinated by science and technology. She dreamed of becoming a researcher and making a difference in the world through her work. One day, while reading a magazine, she came across an article about a scientist named Max and his groundbreaking work in the field of DLAM.

Aabha was immediately intrigued by Max and his work, and she began to research more about DLAM and its potential. She was amazed by the versatility and effectiveness of this technology and the impact it was having on various industries.

Finally, Aabha had the opportunity to meet Max in person at a conference in India. She was star-struck, but she mustered the courage to approach him and ask him about his work. Max was more than happy to answer Aabha's questions, and he was impressed by her knowledge and passion for science.

During their conversation, Max spoke about the importance of collaboration and unity in the field of science and technology. He emphasized that by working together and sharing knowledge, researchers and scientists can achieve more than they ever could alone. This message resonated deeply with Aabha, and she was inspired by Max's belief in the power of collaboration.

From that day on, Aabha was determined to follow in Max's footsteps and make a positive impact on the world through her work. She knew that by fostering a spirit of unity and collaboration, she too could contribute to the advancement of science and technology. And so, Aabha's journey as a young scientist began, all thanks to the significance of DLAM in shaping the future of science and innovation.

In conclusion, DLAM has the potential to shape the future of science and innovation in profound ways. Its versatility, effectiveness, and impact on society make it a technology to watch, and its significance in shaping the future cannot be overstated. It is our responsibility to ensure that DLAM is used for the greater good, and to continue pushing the boundaries of what is possible with this powerful technology.

As we have seen in chapter 22, the significance of DLAM in shaping the future of science and innovation lies in its ability to bring people together in a spirit of unity. With the power to bring together experts from different fields, DLAM has the potential to drive innovation and progress in an inclusive and equitable manner. The potential impact of DLAM is not limited to the field of science and technology, it extends far beyond that, encompassing all aspects of society. In chapter 23, we will delve deeper into the importance of embracing DLAM in finding solutions to the world's problems.

It is evident that the world is facing a multitude of challenges, from global warming to economic disparities. The problems we face are complex and require a collective effort to find solutions. The spirit of unity that DLAM fosters has the power to bring people together, to bridge the divide and to find solutions that are inclusive and equitable. By embracing DLAM, we can take a step forward in the right direction and create a better future for all.

Chapter Twenty-Three

The Importance of Embracing DLAM in Finding Solutions to the World's Problems

In today's world, there are many challenges that are facing humanity, from climate change and the depletion of natural resources, to the spread of disease and poverty. These challenges are complex and often require interdisciplinary approaches to find solutions.

DLAM technology has the potential to be a powerful tool in addressing these global problems. Its ability to process and analyze large amounts of data, make predictions and provide insightful analysis can help to inform decision-making and drive progress in areas such as energy and environmental management, healthcare, and social development.

One example of the potential of DLAM in addressing global problems is its application in healthcare. In a rural area in Africa, a healthcare organization was struggling with limited resources and outdated systems. They were unable to keep up with the growing demand for healthcare services and the data needed to inform their decision-making. With the implementation of DLAM technology, the healthcare organization was able to process and analyze vast amounts of patient data, including medical histories, current health conditions, and demographic information. This allowed them to make informed decisions on how best to allocate their limited resources, prioritize patient care, and improve patient outcomes.

DLAM technology also enabled the organization to predict potential outbreaks of diseases and provide early warnings to local communities. This helped to prevent the spread of diseases and ensured that the most vulnerable populations received the necessary care. Furthermore, DLAM technology allowed the healthcare organization to monitor and track the progress of patients over time, leading to improved patient outcomes and reduced healthcare costs.

This case study highlights the importance of embracing DLAM technology in finding solutions to the world's problems. By processing and analyzing data in real-time, DLAM has the potential to inform decision-making and drive progress in healthcare and other critical areas. In the future, DLAM technology will continue to play a crucial role in addressing some of the world's most pressing problems and improving the lives of people around the globe.

However, in order for DLAM to make its full impact, it is essential that we embrace this technology and work together to unlock its potential. This means working across fields and disciplines to collaborate and share knowledge and expertise, and to continuously strive for improvement and innovation.

Max and Aabha's journey with DLAM technology was nothing short of remarkable. They started off as two skeptical individuals with differing perspectives on the potential of DLAM. However, as they explored the technology further, they realized that DLAM had the potential to revolutionize their respective fields and make a real difference in the world.

Their skepticism soon turned into a passion for DLAM as they began collaborating with experts from different fields, sharing their knowledge and expertise, and learning from each other. They worked tirelessly, developing new applications and use cases for DLAM that had a significant impact in their fields.

Their projects received recognition and awards, and their work was featured in leading industry publications. They became thought leaders in their respective fields, inspiring others to embrace DLAM and work together to find solutions to the world's problems.

However, what set Max and Aabha apart was their unwavering faith in Jesus. They were grateful for all that He had taught them and how their lives had prospered as a result of their faith. They saw DLAM as an opportunity to use their skills and knowledge for a greater purpose, to serve others and make a positive impact on the world.

In the end, Max and Aabha's work with DLAM brought them both great success and fulfillment. They were grateful for the impact that DLAM had on their lives and careers, and for the opportunity to work together and make a positive impact on the world.

Their story is a testament to the importance of embracing DLAM in finding solutions to the world's problems. By leveraging its capabilities and working together in a spirit of unity, we have the opportunity to drive progress and create positive change on a global scale. The future is bright for DLAM, and by working together, we can ensure that its benefits are felt by society for generations to come.

Max and Aabha's story is a testament to the power of DLAM technology and the importance of embracing it in finding solutions to the world's problems. They understood that DLAM had the potential to bring people together and drive progress in a way that was inclusive and equitable, and they were determined to use their skills and knowledge to make a positive impact on the world.

Their story is a testament to the importance of embracing DLAM in finding solutions to the world's problems. By leveraging its capabilities and working together in a spirit of unity, we have the opportunity to drive progress and create positive change on a global scale. The future is bright for DLAM, and by working together, we can ensure that its benefits are felt by society for generations to come.

Another example of collaboration and unity in the Bible can be seen in the story of Nehemiah. Nehemiah was a man who was deeply concerned about the plight of his people and the destruction of the city of Jerusalem. Despite facing opposition from those who sought to stop him, Nehemiah rallied the people together and worked tirelessly to rebuild the walls of the city.

The first chapter of the Book of Nehemiah introduces Nehemiah as a resident of Susa in the Persian Empire. When he heard that the walls of Jerusalem were still broken down despite the temple being rebuilt, he sat down, wept, and fasted while praying to God. The connection between the temple and the wall is significant as they both hold religious and secular importance and were necessary to fulfill God's plan to restore Israel. The temple was dependent on the wall for protection and the wall was dependent on the temple for law, government, security, and prosperity. Nehemiah, who was a trusted advisor and high-ranking official as the king's cupbearer, used his position to lead the rebuilding project and asked for the king's protection for his journey. The text does not suggest there was anything wrong with Nehemiah's decision to seek protection and saw it as evidence of God's blessing, but his approach to trust in God was different from Ezra's, who believed in trusting God without seeking protection.

Through his leadership and determination, Nehemiah was able to bring people from different backgrounds and walks of life together to work towards a common goal. Despite their differences, the people of Jerusalem came together in a spirit of unity and cooperation, using their skills and knowledge to build a better future for themselves and their children.

The story of Nehemiah is a powerful example of what can be achieved when people come together and work towards a common goal. By embracing unity and collaboration, Nehemiah was able to overcome great obstacles and achieve something truly remarkable, leaving a lasting legacy for generations to come. This story demonstrates the importance of working together to find solutions to the world's problems and create positive change for the future.

Max and Aabha were two individuals from different parts of the world who shared a strong faith in Jesus. They both believed that their skills and knowledge were gifts from God, and they desired to use them for a greater purpose.

One day, they heard about the work of a man named Nehemiah who, through his faith in God, led the rebuilding of the walls of Jerusalem and brought the people of Israel together in unity.

This inspired Max and Aabha to take on a similar challenge, using their skills and knowledge to help rebuild and improve their respective communities. They saw DLAM technology as a tool that could help them achieve their goal.

Max, a computer engineer, used DLAM to analyze data and provide insights on how to improve infrastructure and services in underprivileged areas. Aabha, a healthcare professional, used DLAM to identify areas in need of improvement in healthcare access and delivery.

Together, they worked with local leaders and communities to bring about positive change, just as Nehemiah had done in Jerusalem. They embraced the spirit of unity and collaboration, and through their hard work, they were able to make a significant impact on the lives of those around them.

Their efforts received recognition and praise, and their work inspired others to join in the mission of rebuilding and improving their communities. Max and Aabha realized that through their faith in Jesus and their embrace of DLAM technology, they had become today's Nehemiahs, bringing people together in unity to achieve a common goal.

Their journey showed that, with faith and determination, it is possible to bring about positive change and make a lasting impact on the world.

Afterword

As I reflect on the publication of "IR Physics 2 - Innovations in AI Safety and Longevity Through IR Physics and DLAM," I can't help but marvel at the progress we have made as a society. The concepts and theories discussed in this book were once the stuff of science fiction, but now they are a reality.

The field of IR Physics has continued to evolve and expand, with new discoveries and applications being made every day. The potential for DLAM to revolutionize energy production and solve some of the world's most pressing problems is truly exciting.

As we look to the future, it is clear that the principles of IR Physics and DLAM will play an increasingly important role in our lives. From sustainable energy production to medical breakthroughs, these concepts hold the key to a brighter future for us all.

But with this potential comes responsibility. We must ensure that the development and application of these technologies are done in a responsible and ethical manner, with a focus on the well-being of individuals and communities.

I hope that the discussions and ideas presented in this book inspire future generations to continue pushing the boundaries of what is possible. Let us work together to create a world where science and technology are harnessed for the betterment of all. The future is bright, and with the principles of IR Physics and DLAM guiding us, we can achieve anything we set our minds to.

About Author

Meet the author who champions the underdogs and brings hidden innovation to the forefront. As someone who is Least Commanded (Matthew 5:19), I strive to create positive change by protecting the vulnerable in complex situations. My expertise in creating Child Safety Files for separated parents through MyReal BestToolChest.com, managing organic data metrics, and working with families and nonprofits has earned recognition from news leaders, book reviewers, and government organizations. Guided by a passion for justice and equality, I am committed to building a brighter and safer future for all. So, let's safeguard our dreams and work towards a better world together!

Also By

<u>Change the Game: One Simple New Deal Can Help Any Family Change Any Industry (Series)</u>

A Light in the Dark: Ace and Crush's Story of Courage and Determination

Adorning Your Permanent Losses So the Future You Adores It (CLU Self-Help Guide)

Breaking the Mold: Transforming Legal Learning with Professional Law Caveats - (Family Law 2.0 Second Edition)

Dealtown Tailoring: Surviving Anguish or Success in America?

Faithfinder: Inclusive Holiday Stories Empower Readers to Find Their Super-faith!

Family Law 2.0: Hatching a Superstate (Family Law 2.0 First Edition)

Impossible Data: How Making Almighty Data Cavityism History Can Help Us Live A Better Life

Impossible Psychologies

IR Physics 2

Parenting 2.0 For Un-Haters

Revitalizing Impossible Brothers: Award Yourself Kids Making Sibling Cavityism History

Revitalizing Education: The Role of Academic Law Caveats in Student Growth

Revitalizing Inclusive Socioeconomics: The Role of SocioEcTech Justice on Human Growth

Outside The Series:

American Pros

Family Law 2.0

I R "Freestyle" Physics – The Study of of Transformation and/or Transfiguration

Law & Side-Law: Family Hostility / Family Empathy

The Law Doctor

Uphright USA

United CLU Parents Guide

Made in the USA
Monee, IL
04 November 2026

67377400R00079